Linear Algebra for Calculus

Konrad J. Heuvers
John H. Kuisti
William P. Francis
Gene M. Ortner
Daniel S. Moak
Deborah F. Lockhart

Michigan Technological University

Brooks/Cole Publishing Company
Pacific Grove, California

Brooks/Cole Publishing Company
A Division of Wadsworth, Inc.

Printed in the United States of America

10 9 8 7 6 5 4 3 2 1

ISBN 0-534-15402-6

Sponsoring Editor: Faith B. Stoddard
Editorial Assistant: Nancy Champlin
Production Coordinator: Dorothy Bell
Cover Design: Katherine Minerva
Cover Photo: Lee Hocker
Printing and Binding: Malloy Lithographing, Inc.

Preface

This book was designed to be used as a supplement to the calculus course at Michigan Technological University. It provides a quick introduction to the basic ideas of linear algebra. The emphasis of the first five chapters is on fundamental matrix operations and concepts, solutions of systems of equations, determinants, and inverses. Chapters 6, 7, and 8 cover orthogonal matrices, changes of basis, the eigenvalue problem, and applications of the eigenvalue problem to the analysis of quadratic equations in two and three variables. This material assumes a knowledge of two and three dimensional vector algebra. Chapter 9 is a set of miscellaneous problems for the first eight chapters. Chapters 10, 11, 12, and 13 cover the chain rule in matrix form, linear and quadratic approximations, the use of eigenvalues to analyze critical points for functions of several variables, and the eigenvalue problem as a special case of Lagrange multipliers with quadratic functions and constraints. This material can be covered in a multivariate calculus course or after such a course.

AT MTU we use the first eight chapters of the book in our calculus courses after vector algebra has been covered. The remaining material is used by some of the authors in their multivariate calculus courses.

We would like to especially acknowledge the advice, comments, and suggestions that Clark R. Givens, MTU, has given in the development of this material. We would also like to thank our colleagues Thomas Drummer, Lee W. Erlebach, Sidney Graham, John W. Hilgers, John W. Kern, Kenneth L. Kuttler, and Tom Martin for their helpful suggestions.

Michigan Technological University
Houghton, Michigan

William P. Francis
Konrad J. Heuvers
John H. Kuisti
Deborah F. Lockhart
Daniel S. Moak
Gene M. Ortner

Contents

1 Matrices and Matrix Algebra

Outline of Key Ideas

Use of matrices in the representation of systems of linear equations.

Definitions:

matrix, element, row, column

equality

zero matrix

addition, subtraction

scalar multiplication of a matrix

matrix multiplication

diagonal matrix

identity matrix

transpose.

Algebraic Properties of Matrices.

Matrices and Matrix Algebra

Many problems in engineering, business, and the sciences involve the study of an appropriate system of linear equations.

Example 1.1 The system

$$\begin{aligned} 2x_1 - 3x_2 + 4x_3 &= 0 \\ x_1 + x_2 + x_3 &= 6 \end{aligned} \tag{1.1}$$

consists of two equations and involves three unknowns.

We shall see that it is very useful to represent the system (1.1) using rectangular arrays of numbers or symbols. Such arrays are called ***matrices***. Using matrices system (1.1) can be written as

$$A\mathbf{x} = \mathbf{b} \tag{1.2}$$

where A is the ***coefficient matrix*** which is given by

$$A = \begin{pmatrix} 2 & -3 & 4 \\ 1 & 1 & 1 \end{pmatrix} \tag{1.3}$$

and where $\mathbf{x}$ and $\mathbf{b}$ are ***column matrices*** (also called ***column vectors***) which are given by

$$\mathbf{x} = \begin{pmatrix} x_1 \\ x_2 \\ x_3 \end{pmatrix} \quad \text{and} \quad \mathbf{b} = \begin{pmatrix} 0 \\ 6 \end{pmatrix} . \tag{1.4}$$

In the middle of the 19th century Arthur Cayley, an English mathematician, realized that the study of equations could be greatly facilitated by studying their matrices as separate algebraic objects. In this lesson we will introduce the elementary operations and concepts of the matrix algebra developed by Cayley.

1.1 A Table of Definitions

	Definition	Example
1.	A rectangular array of numbers or symbols is called a ***matrix***. The numbers or symbols in the matrix are called its ***entries***.	$\begin{pmatrix} 1 & 2 & 3 & 4 \\ 5 & 0 & 1 & 2 \\ 3 & 2 & 1 & 0 \end{pmatrix}$ This matrix has 12 entries.

2. The sequence of all entries on a horizontal line is called a ***row*** . The rows are numbered from top to bottom.

$$\begin{pmatrix} 1 & 2 & 3 & 4 \\ \mathbf{5} & \mathbf{0} & \mathbf{1} & \mathbf{2} \\ 3 & 2 & 1 & 0 \end{pmatrix}$$

The second row is in **boldface type**.

3. The set of all entries on a vertical line is called a ***column*** . The columns are numbered from left to right.

$$\begin{pmatrix} 1 & 2 & \mathbf{3} & 4 \\ 5 & 0 & \mathbf{1} & 2 \\ 3 & 2 & \mathbf{1} & 0 \end{pmatrix}$$

The third column is in **boldface type**.

4. The entry in the ith row and in the jth column is in the (i,j) ***position***, is often represented by a doubly-subscripted symbol such as a_{ij}, and is referred to as the (i,j) ***entry***.

$$\begin{pmatrix} 1 & 2 & \mathbf{3} & 4 \\ 5 & 0 & 1 & 2 \\ 3 & 2 & 1 & 0 \end{pmatrix}$$

The $(1,3)$ entry is printed in **boldface type**. We have $a_{13} = 3$.

5. A matrix with m rows and n columns is called an $m \times n$ ***matrix***. Often we will indicate a matrix as: A is the $m \times n$ matrix $[a_{ij}]$.

$$A = \begin{pmatrix} a_{11} & a_{12} & a_{13} \\ a_{21} & a_{22} & a_{23} \end{pmatrix}$$

A is the 2×3 matrix $[a_{ij}]$.

6. A matrix with the same number of rows as columns is called a ***square matrix***.

$$\begin{pmatrix} 1 & 2 & 3 \\ 1 & 2 & 2 \\ 1 & 1 & 1 \end{pmatrix}$$

This is a 3×3 square matrix.

7. A $1 \times n$ matrix is called a ***row matrix*** or ***row vector***.

$$\begin{pmatrix} 1 & 2 & 3 & 4 \end{pmatrix}$$

This is 1×4 row vector.

8. An $m \times 1$ matrix is called a ***column matrix*** or ***column vector***.

$$\begin{pmatrix} 1 \\ 2 \\ 3 \end{pmatrix}$$

This is a 3×1 column vector.

9. **Matrix Equality**

Given two $m \times n$ matrices, $A = [a_{ij}]$ and $B = [b_{ij}]$, then $A = B$ if and only if $a_{ij} = b_{ij}$ for $i = 1, 2, \ldots, m$ and $j = 1, 2, \ldots, n$

$$\begin{pmatrix} x_{11} & x_{12} \\ x_{21} & x_{22} \end{pmatrix} = \begin{pmatrix} 1 & 2 \\ 3 & 4 \end{pmatrix}$$

if and only if $x_{11} = 1$, $x_{12} = 2$, $x_{21} = 3$, and $x_{22} = 4$.

10. **Matrix Addition**

If $A = [a_{ij}]$ and $B = [b_{ij}]$ are both $m \times n$, then $A + B$ is the $m \times n$ matrix $[a_{ij} + b_{ij}]$.

$$\begin{pmatrix} 2 & 1 \\ 3 & 2 \end{pmatrix} + \begin{pmatrix} 4 & 2 \\ 1 & 5 \end{pmatrix} = \begin{pmatrix} 6 & 3 \\ 4 & 7 \end{pmatrix}$$

11. ***Scalar multiplication*** of a matrix $A = [a_{ij}]$ by any number α means $\alpha A = [\alpha a_{ij}]$.

$$2\begin{pmatrix} 3 & 2 \\ 1 & 4 \end{pmatrix} = \begin{pmatrix} 6 & 4 \\ 2 & 8 \end{pmatrix}$$

12. **Zero Matrices**

$\mathcal{O} = [z_{ij}]$ where $z_{ij} = 0$ for $i = 1, 2, \ldots, m$ and $j = 1, 2, \ldots, n$

$$\mathcal{O} = \begin{pmatrix} 0 & 0 \\ 0 & 0 \end{pmatrix}, \qquad \mathcal{O} = \begin{pmatrix} 0 & 0 \\ 0 & 0 \\ 0 & 0 \end{pmatrix}$$

13. **Matrix Subtraction**

If A and B are both $m \times n$ matrices, then $A - B = A + (-1)B$.

$$\begin{pmatrix} 3 & 2 \\ 4 & 1 \end{pmatrix} - \begin{pmatrix} 2 & 1 \\ 0 & 1 \end{pmatrix} =$$
$$\begin{pmatrix} 3 & 2 \\ 4 & 1 \end{pmatrix} + \begin{pmatrix} -2 & -1 \\ 0 & -1 \end{pmatrix} = \begin{pmatrix} 1 & 1 \\ 4 & 0 \end{pmatrix}$$

14. Sometimes we will indicate an $m \times n$ matrix A by writing

$$A = \begin{pmatrix} \mathbf{a}_{(1)} \\ \mathbf{a}_{(2)} \\ \vdots \\ \mathbf{a}_{(m)} \end{pmatrix}$$

where $\mathbf{a}_{(i)}$ represents the ith row of A or by

$$A = \begin{pmatrix} \mathbf{a}_1 & \mathbf{a}_2 & \cdots & \mathbf{a}_n \end{pmatrix}$$

where $\mathbf{a}_j$ represents the jth column of A.

$$A = \begin{pmatrix} 1 & 3 & 2 \\ 5 & 6 & 1 \end{pmatrix} = \begin{pmatrix} \mathbf{a}_{(1)} \\ \mathbf{a}_{(2)} \end{pmatrix}$$
$$= \begin{pmatrix} \mathbf{a}_1 & \mathbf{a}_2 & \mathbf{a}_3 \end{pmatrix}$$

where, for example, $\mathbf{a}_{(2)} = \begin{pmatrix} 5 & 6 & 1 \end{pmatrix}$

and $\mathbf{a}_3 = \begin{pmatrix} 2 \\ 1 \end{pmatrix}$

15. If the number of columns in matrix A is equal to the number of rows in matrix B, we say A A ***is conformable with respect to*** B .

$$C = \begin{pmatrix} 1 & 2 \\ 3 & 4 \\ 5 & 6 \end{pmatrix} \qquad D = \begin{pmatrix} 1 & 0 & 1 & 3 \\ 1 & 1 & 1 & 2 \end{pmatrix}$$

C is 3×2 and D is 2×4. C is conformable with respect to D, but D is ***not*** conformable with respect to C.

16. If A is a $1 \times n$ row matrix

$$A = \begin{pmatrix} a_1 & a_2 & \cdots & a_n \end{pmatrix}$$

and B is a $n \times 1$ column matrix

$$B = \begin{pmatrix} b_1 \\ b_2 \\ \vdots \\ b_n \end{pmatrix}$$

(hence A is *conformable with respect to* B), we define AB as the 1×1 matrix $\begin{pmatrix} a_1b_1 + a_2b_2 + \cdots + a_nb_n \end{pmatrix}$. This is a row times column multiplication. Note that in vector language this is just the dot product of the row vector A and the column vector B.

$$\begin{pmatrix} 1 & 2 & 3 \end{pmatrix} \begin{pmatrix} 4 \\ 5 \\ 6 \end{pmatrix} =$$

$$(1 \cdot 4 + 2 \cdot 5 + 3 \cdot 6) = (32)$$

17. **Matrix Multiplication**

If A is $m \times n$ and B is $n \times s$ (hence A is *conformable with respect to* B), then we define $P = AB$ as the $m \times s$ matrix $[p_{ij}]$ where $p_{ij} = \mathbf{a}_{(i)} \bullet \mathbf{b}_j$ is the result of multiplying the ith row of A times the jth column of B. Note that ms multiplications must be performed to obtain the entire $m \times s$ matrix P.

$$\underset{2 \times 3}{\begin{pmatrix} 1 & 2 & 3 \\ 3 & 4 & 0 \end{pmatrix}} \underset{3 \times 2}{\begin{pmatrix} 1 & 1 \\ 0 & 1 \\ 1 & 2 \end{pmatrix}} =$$

$$\underset{2 \times 2}{\begin{pmatrix} 4 & 9 \\ 3 & 7 \end{pmatrix}}$$

. Note that $p_{21} = \mathbf{a}_{(2)} \bullet \mathbf{b}_1 =$

$$\begin{pmatrix} 3 & 4 & 0 \end{pmatrix} \begin{pmatrix} 1 \\ 0 \\ 1 \end{pmatrix} = 3 + 0 + 0 = 3.$$

18. If A is a square matrix, then the entries a_{ii} of A with equal row and column indices form the ***diagonal*** of A.

$$\begin{pmatrix} \mathbf{1} & 2 & 3 \\ 4 & \mathbf{5} & 6 \\ 7 & 8 & \mathbf{9} \end{pmatrix}$$

The diagonal is given in **boldface type**.

19. A square matrix which has all non-diagonal entries equal to zero is called a ***diagonal matrix***.

$$\begin{pmatrix} 1 & 0 & 0 \\ 0 & 2 & 0 \\ 0 & 0 & 3 \end{pmatrix}$$

20. The $n \times n$ diagonal matrix with every diagonal entry equal to 1 is called the $n \times n$ ***identity matrix*** I_n.

$$I_4 = \begin{pmatrix} 1 & 0 & 0 & 0 \\ 0 & 1 & 0 & 0 \\ 0 & 0 & 1 & 0 \\ 0 & 0 & 0 & 1 \end{pmatrix}$$

21. If $A = [a_{ij}]$ is an $m \times p$ matrix and $B = [b_{ij}]$ is a $p \times m$ matrix and if $a_{ij} = b_{ji}$ for $i = 1, 2, \ldots, m$ and $j = 1, 2, \ldots, p$, then B is called the ***transpose*** of A. The symbol A^T is often used for B. Note that the entries of the ith row of A become the entries of the ith column of A^T.

$$A = \begin{pmatrix} 1 & 0 & 2 \\ 3 & 1 & 1 \end{pmatrix}$$

$$A^T = \begin{pmatrix} 1 & 3 \\ 0 & 1 \\ 2 & 1 \end{pmatrix}$$

$$C = \begin{pmatrix} 1 & 2 & 3 \end{pmatrix}$$

$$C^T = \begin{pmatrix} 1 \\ 2 \\ 3 \end{pmatrix}$$

1.2 Algebraic Properties of Matrices

The algebraic properties of matrix addition and subtraction, matrix multiplication, and scalar multiplication of matrices are similar to those of the real numbers. These properties are listed below. A, B, C, I, and $\mathcal{O}$ are matrices and α is a number. When a matrix product such as AB appears, it is assumed that A is conformable with respect to B.

$$A + \mathcal{O} = \mathcal{O} + A = A \tag{1.5}$$

$$A\mathcal{O} = \mathcal{O}A = \mathcal{O} \tag{1.6}$$

$$AI = IA = A \tag{1.7}$$

$$A - A = \mathcal{O} \tag{1.8}$$

$$A + B = B + A \tag{1.9}$$

$$A(B + C) = AB + AC \tag{1.10}$$

$$(B + C)A = BA + CA \tag{1.11}$$

$$(A + B) + C = A + (B + C) \tag{1.12}$$

$$(AB)C = A(BC) \tag{1.13}$$

$$\alpha(AB) = (\alpha A)B = A(\alpha B) \tag{1.14}$$

$$(AB)^T = B^T A^T \; . \tag{1.15}$$

If this were a list of properties for real numbers, the commutative property of multiplication, $ab = ba$ would appear. It is most important to realize that ***matrix multiplication is not generally commutative***! The product of two specific matrices may happen to commute (such as in equations (1.6) and (1.7) above) but that is not usually the case.

Example 1.2 A is a 3×2 matrix and B is a 2×5 matrix. A is conformable with respect to B so that AB may be obtained. But B is ***not*** conformable with respect to A so that BA is not even defined. Hence AB exists but BA does not exist.

Example 1.3 A is a 3×2 matrix and B is a 2×3 matrix. Then both AB and BA are defined, but AB is a 3×3 matrix while BA is a 2×2 matrix. Hence it is not possible for AB and BA to be equal matrices.

Example 1.4 Even if A and B are square $n \times n$ matrices, then matrix multiplication still need not be commutative. For example, let

$$A = \begin{pmatrix} 3 & 6 \\ -4 & -8 \end{pmatrix}$$

and

$$B = \begin{pmatrix} -10 & -4 \\ 5 & 2 \end{pmatrix} \quad .$$

Then

$$AB = \begin{pmatrix} 0 & 0 \\ 0 & 0 \end{pmatrix}$$

and

$$BA = \begin{pmatrix} -14 & -28 \\ 7 & 14 \end{pmatrix}$$

so AB is not equal to BA!

With Definition 9 of matrix equality and Definition 17 matrix multiplication, you should be able to verify that matrix Equation (1.2) with the matrices defined in (1.3) and (1.4) is an equivalent way of writing the system of equations (1.1).

Another use of matrix algebra was mentioned in Definitions 16 and 17 and is stated more explicitly in the next example.

Example 1.5 Represent the dot product of two vectors in matrix language.

Solution.

Suppose the two vectors each have n components:

$$a = \langle a_1, a_2, \ldots, a_n \rangle \qquad b = \langle b_1, b_2, \ldots, b_n \rangle \quad .$$

We shall ***choose*** to represent each vector by an $n \times 1$ column matrix whose entries are the components of the vector, so we write

$$\mathbf{a} = \begin{pmatrix} a_1 \\ a_2 \\ \vdots \\ a_n \end{pmatrix} \qquad \mathbf{b} = \begin{pmatrix} b_1 \\ b_2 \\ \vdots \\ b_n \end{pmatrix} .$$

Notice that $\mathbf{a}^T$ and $\mathbf{b}^T$, the transposes of $\mathbf{a}$ and $\mathbf{b}$, are $1 \times n$ row matrices. Then the definitions of matrix multiplication and the dot product of vectors permit the following statements:

$$\begin{aligned} \mathbf{a}^T\mathbf{b} &= a_1b_1 + a_2b_2 + \cdots + a_nb_n = \mathbf{a} \bullet \mathbf{b} = \\ &= \mathbf{b} \bullet \mathbf{a} = b_1a_1 + b_2a_2 + \cdots + b_na_n = \mathbf{b}^T\mathbf{a} . \end{aligned}$$

1.3 Exercises

1. Compute

$$2\begin{pmatrix} 1 & 2 & 5 \\ 3 & 4 & 0 \end{pmatrix} - 3\begin{pmatrix} 0 & 2 & 1 \\ 1 & 4 & -3 \end{pmatrix} .$$

2. Compute

$$\begin{pmatrix} 1 & 2 & 1 \\ -1 & 6 & 4 \\ 3 & 1 & 2 \end{pmatrix}\begin{pmatrix} 1 & 1 \\ 0 & -1 \\ 2 & 3 \end{pmatrix} .$$

3. Compute

$$\begin{pmatrix} 3 & 0 & 0 \\ 0 & 3 & 0 \\ 0 & 0 & 3 \end{pmatrix}\begin{pmatrix} 2 & 1 & 1 & 5 \\ 1 & 0 & 1 & 3 \\ 2 & 4 & 1 & 5 \end{pmatrix} .$$

4. Compute

$$\begin{pmatrix} \alpha & 0 & 0 \\ 0 & \alpha & 0 \\ 0 & 0 & \alpha \end{pmatrix}\begin{pmatrix} 1 & 6 & 5 \\ 3 & 0 & 1 \\ 2 & 1 & 1 \end{pmatrix}$$

where α is any real number.

5. Compute

$$\begin{pmatrix} \alpha & 0 \\ 0 & \beta \end{pmatrix} \begin{pmatrix} 1 & 5 & 3 & 2 \\ 0 & 1 & 1 & 1 \end{pmatrix}$$

where α and β are real numbers.

6. Show

$$\left(\begin{pmatrix} 3 & 1 \\ -2 & 1 \end{pmatrix} \begin{pmatrix} 1 & 1 \\ 2 & 1 \end{pmatrix} \right) + \left(\begin{pmatrix} 3 & 1 \\ -2 & 1 \end{pmatrix} \begin{pmatrix} 1 & -1 \\ -2 & -1 \end{pmatrix} \right) =$$

$$\begin{pmatrix} 3 & 1 \\ -2 & 1 \end{pmatrix} \left(\begin{pmatrix} 1 & 1 \\ 2 & 1 \end{pmatrix} + \begin{pmatrix} 1 & -1 \\ -2 & -1 \end{pmatrix} \right)$$

by computing each side separately.

7. Compute AI_3 and I_2A where

$$A = \begin{pmatrix} 3 & 2 & 5 \\ -2 & 4 & 6 \end{pmatrix} .$$

8. Show

$$\left(\begin{pmatrix} 1 & 3 & 6 \\ 4 & 2 & 1 \end{pmatrix} \begin{pmatrix} 3 \\ 0 \\ 2 \end{pmatrix} \right)^T = \begin{pmatrix} 3 & 0 & 2 \end{pmatrix} \begin{pmatrix} 1 & 4 \\ 3 & 2 \\ 6 & 1 \end{pmatrix}$$

by computing each side separately.

9. If

$$\begin{pmatrix} a & b \\ c & d \end{pmatrix} \begin{pmatrix} -11 & -3 \\ 4 & 1 \end{pmatrix} = \begin{pmatrix} 1 & 0 \\ 0 & 1 \end{pmatrix} ,$$

find a, b, c, and d.

10. If

$$\begin{aligned} y &= 2w + x \\ z &= w + x \end{aligned}$$

and

$$\begin{array}{l} u = 4y \\ v = 3y - 4z \end{array},$$

then write each system as a matrix equation and write a matrix equation relating

$$\begin{pmatrix} u \\ v \end{pmatrix} \quad \text{and} \quad \begin{pmatrix} w \\ x \end{pmatrix}.$$

11. Find the following for the matrix

$$A = \begin{pmatrix} 1 & -1 & 2 & 3 \\ 0 & 2 & -3 & 1 \\ 5 & 3 & -1 & 2 \end{pmatrix}$$

a) a_{32} **b)** a_{23} **c)** the $m \times n$ dimensions of A

d) $\mathbf{a}_2$ where $A = \begin{pmatrix} \mathbf{a}_1 & \mathbf{a}_2 & \mathbf{a}_3 & \mathbf{a}_4 \end{pmatrix}$ **e)** $\mathbf{a}_{(3)}$ where $A = \begin{pmatrix} \mathbf{a}_{(1)} \\ \mathbf{a}_{(2)} \\ \mathbf{a}_{(3)} \end{pmatrix}$ **f)** A^T .

12. If $\mathbf{a} = \langle 1, -1, 3 \rangle$ and $\mathbf{b} = \langle 2, 1, -1 \rangle$, write these vectors as column matrices and find $\mathbf{a} \bullet \mathbf{b}$ using matrix multiplication.

13. Suppose that $A = \begin{pmatrix} -1 & 3 & 7 \\ 5 & 1 & 4 \end{pmatrix}$, $B = \begin{pmatrix} 5 & 1 & 11 \\ 9 & 3 & 12 \end{pmatrix}$, and $A + 2X = B$. Find X.

14. Compute the following

a) $\begin{pmatrix} 2 \\ 1 \\ -1 \end{pmatrix} \begin{pmatrix} 1 & 2 & 3 \end{pmatrix} + \begin{pmatrix} 1 \\ -1 \\ 1 \end{pmatrix} \begin{pmatrix} 3 & 2 & 1 \end{pmatrix}$

b) $\begin{pmatrix} 1 & 2 & 3 \end{pmatrix} \begin{pmatrix} 2 \\ 1 \\ -1 \end{pmatrix} + \begin{pmatrix} 3 & 2 & 1 \end{pmatrix} \begin{pmatrix} 1 \\ -1 \\ 1 \end{pmatrix}$.

15. If $(A|B)$ is a partitioned matrix with A in the first part and B in the second part, verify that $P(A|B) = (PA|PB)$ if

a) $P = \begin{pmatrix} 1 & 2 \\ 3 & -1 \end{pmatrix}$ $\quad A = \begin{pmatrix} 1 & 1 \\ 1 & -1 \end{pmatrix}$ $\quad B = \begin{pmatrix} 2 & -1 \\ 1 & 3 \end{pmatrix}$

b) $P = \begin{pmatrix} 1 & 1 & -1 \\ -1 & 2 & 1 \\ 2 & 1 & 1 \end{pmatrix}$ $\quad A = \begin{pmatrix} 1 & 1 & 1 \\ 1 & 1 & 1 \\ 1 & 1 & 1 \end{pmatrix}$ $\quad B = \begin{pmatrix} 0 & 1 & 2 \\ -1 & 0 & 3 \\ -2 & -3 & 0 \end{pmatrix}$.

2 Solving a System of Linear Equations with Elementary Row Operations

Outline of Key Ideas

Matrix representation of a system of linear equations.

Elementary row operations on matrices.

Solving a system of linear equations:

the row reduction method (Gaussian elimination)

row reduced echelon forms.

Solving a System of Linear Equations with Elementary Row Operations

Much terminology and many concepts about matrices and their algebra have been introduced in Chapter 1. Some very important additional manipulations on matrices called the ***elementary row operations*** and their use in solving a system of linear equations will now be studied. These operations will be extensively employed in later lessons to solve a variety of systems of linear equations, to evaluate determinants, and to find inverse matrices.

2.1 Matrix Representation of a System of Linear Equations

The elementary row operations correspond to some simple algebraic manipulations used in solving systems of linear equations. To exploit this connection, consider the matrix representation of such a system presented in Example 2.1.

Example 2.1 The system

(2.1)
$$\begin{aligned} x_1 + 2x_2 + 3x_3 + \ 4x_4 + \ 5x_5 &= 6 \\ 7x_1 + 8x_2 + 9x_3 + 10x_4 + 11x_5 &= 12 \end{aligned}$$

can be expressed in terms of matrices as

(2.2)
$$\begin{pmatrix} 1 & 2 & 3 & 4 & 5 \\ 7 & 8 & 9 & 10 & 11 \end{pmatrix} \begin{pmatrix} x_1 \\ x_2 \\ x_3 \\ x_4 \\ x_5 \end{pmatrix} = \begin{pmatrix} 6 \\ 12 \end{pmatrix} .$$

Note that (2.2) is a single matrix equation of the form

(2.3)
$$A\mathbf{x} = \mathbf{b} .$$

The matrix

(2.4)
$$A = \begin{pmatrix} 1 & 2 & 3 & 4 & 5 \\ 7 & 8 & 9 & 10 & 11 \end{pmatrix}$$

is called the ***coefficient matrix*** of the system (2.1). A solution vector of the system is any column matrix which satisfies Equation (2.2) when substituted for

(2.5)
$$\mathbf{x} = \begin{pmatrix} x_1 \\ x_2 \\ x_3 \\ x_4 \\ x_5 \end{pmatrix} .$$

Another important matrix associated with the system (2.1), the ***augmented matrix*** , is constructed by appending the column matrix $\mathbf{b}$ to the coefficient matrix A. In this example the augmented matrix $[A|\mathbf{b}]$ is

(2.6)
$$[A|\mathbf{b}] = \left(\begin{array}{ccccc|c} 1 & 2 & 3 & 4 & 5 & 6 \\ 7 & 8 & 9 & 10 & 11 & 12 \end{array}\right) .$$

The augmented matrix is a complete record of the specific information given for the system. For example, if the matrix

(2.7)
$$\left(\begin{array}{cc|c} 3 & 2 & 4 \\ 1 & -1 & 3 \end{array}\right)$$

were used as an augmented matrix, it would correspond to the system of equations

(2.8)
$$\begin{aligned} 3x_1 + 2x_2 &= 4 \\ x_1 - x_2 &= 3 \end{aligned} .$$

2.2 Elementary Row Operations on Matrices

Note that the kth row of the augmented matrix records the information contained in the kth equation of the system. An algebraic operation performed upon an equation in the system will become an arithmetical operation performed on the corresponding row of the augmented matrix. We will consider three elementary operations on the system. The corresponding operations on the rows of the augmented matrix are called ***elementary row operations.*** To emphasize the correspondence, the two versions are presented side-by-side.

Operation of Type 1:

Interchange two equations — Interchange two rows

(2.9)
$$\begin{aligned} 2x + y &= 1 \\ 3x - y &= 2 \\ x + y &= -1 \end{aligned} \qquad \begin{aligned} \mathbf{x + y} &= \mathbf{-1} \\ 3x - y &= 2 \\ \mathbf{2x + y} &= \mathbf{1} \end{aligned} \qquad \left(\begin{array}{cc|c} 2 & 1 & 1 \\ 3 & -1 & 2 \\ 1 & 1 & -1 \end{array}\right) \left(\begin{array}{cc|c} \mathbf{1} & \mathbf{1} & \mathbf{-1} \\ 3 & -1 & 2 \\ \mathbf{2} & \mathbf{1} & \mathbf{1} \end{array}\right) .$$

Operation of Type 2:

Multiply one equation by a non-zero number.

Multiply one row by a non-zero number.

(2.10)

$$\begin{aligned} x - y &= 1 \\ 2x + y &= -1 \end{aligned} \qquad \begin{aligned} \mathbf{-2x + 2y} &= \mathbf{-2} \\ 2x + \ \ y &= -1 \end{aligned}$$

$$\left(\begin{array}{cc|c} 1 & -1 & 1 \\ 2 & 1 & -1 \end{array}\right) \left(\begin{array}{cc|c} \mathbf{-2} & \mathbf{2} & \mathbf{-2} \\ 2 & 1 & -1 \end{array}\right).$$

Operation of Type 3:

Replace one equation by itself plus a scalar multiple of another equation.

Replace one row by itself plus a scalar multiple of another row.

(2.11)

$$\begin{aligned} x + \ y &= 1 \\ -x + 2y &= 2 \\ 3x - \ y &= -1 \end{aligned} \qquad \begin{aligned} x + \ y &= 1 \\ -x + 2y &= 2 \\ -\mathbf{4y} &= \mathbf{-4} \end{aligned}$$

$$\left(\begin{array}{cc|c} 1 & 1 & 1 \\ -1 & 2 & 2 \\ 3 & -1 & -1 \end{array}\right) \left(\begin{array}{cc|c} 1 & 1 & 1 \\ -1 & 2 & 2 \\ \mathbf{0} & \mathbf{-4} & \mathbf{-4} \end{array}\right).$$

Note that -3 times the first equation has been added to the third equation. As a result of this operation, the third equation is ***replaced*** while the first equation in unaffected! Also note that no other equation is involved.

2.3 Solving a System of Linear Equations

The application of any one of these operations to a system of linear equations

$$A_1\mathbf{x} = \mathbf{b}_1 \quad \{\text{with augmented matrix } [A_1|\mathbf{b}_1]\,\}$$

produces another system

$$A_2\mathbf{x} = \mathbf{b}_2 \quad \{\text{with augmented matrix } [A_2|\mathbf{b}_2]\,\} \quad .$$

The important connection between these two systems is that every solution vector $\mathbf{x}$ of one system is also a solution vector of the other system, i.e. the two systems have the same ***solution set.*** Such systems are said to be ***equivalent*** .

Since a sequence of elementary operations transforms a system into a sequence of equivalent systems, we have the following strategy available to solve a system of linear equations.

Apply such a sequence to the system to obtain another system which is easier to solve. Find its solution set. This will also be the solution set of the original system since the two systems are equivalent.

Example 2.2 Find the solution set of the system

$$(2.12) \qquad \begin{aligned} 3x + 2y &= 4 \\ x - y &= 3 \end{aligned} \qquad \left(\begin{array}{rr|r} 3 & 2 & 4 \\ 1 & -1 & 3 \end{array}\right).$$

Solution.

Apply an operation of type 1 (*interchange the two equations*) to obtain

$$(2.13) \qquad \begin{aligned} x - y &= 3 \\ 3x + 2y &= 4 \end{aligned} \qquad \left(\begin{array}{rr|r} 1 & -1 & 3 \\ 3 & 2 & 4 \end{array}\right).$$

Apply an operation of type 3 (*−3 times the first equation added to the second equation*) to obtain

$$(2.14) \qquad \begin{aligned} x - y &= 3 \\ 5y &= -5 \end{aligned} \qquad \left(\begin{array}{rr|r} 1 & -1 & 3 \\ 0 & 5 & -5 \end{array}\right).$$

Apply an operation of type 2 (*1/5 times the second equation*) to obtain

$$(2.15) \qquad \begin{aligned} x - y &= 3 \\ y &= -1 \end{aligned} \qquad \left(\begin{array}{rr|r} 1 & -1 & 3 \\ 0 & 1 & -1 \end{array}\right).$$

Apply an operation of type 3 (*1 times the second equation added to the first equation*) to obtain

$$(2.16) \qquad \begin{aligned} x &= 2 \\ y &= -1 \end{aligned} \qquad \left(\begin{array}{rr|r} 1 & 0 & 2 \\ 0 & 1 & -1 \end{array}\right).$$

This system clearly has only one solution $\begin{pmatrix} 2 \\ -1 \end{pmatrix}$. The reader should check that this solution vector does indeed satisfy the original system (2.12).

2.4 The Row Reduction Method (Gaussian Elimination)

If an elementary row operation is applied to a given matrix to produce another matrix, the two matrices are said to be ***row equivalent*** (*whether or not they are playing the role of augmented matrices*). Hence each augmented matrix in the sequence in the example is row equivalent to every other augmented matrix in the sequence. Two row equivalent augmented matrices represent two equivalent systems of equations. The ***simplest*** system to solve has an augmented matrix $[A_0|\mathbf{b}_0]$ called the ***row-reduced echelon form*** of the original system's augmented matrix $[A|\mathbf{b}]$. In the example, the augmented matrix (2.16) is the row-reduced echelon form of (2.12).

The method of solving systems of linear equations demonstrated in the example is called ***Gaussian Elimination***. Using the language of augmented matrices, it can be succinctly described as follows:

1) Write the augmented matrix $[A|\mathbf{b}]$ for the given system

2) Find its row-reduced echelon form $[A_0|\mathbf{b}_0]$

3) Solve the system of equations corresponding to the augmented matrix $[A_0|\mathbf{b}_0]$.

2.5 Row-Reduced Echelon Forms

Now how does one recognize a row-reduced echelon form? A non-zero $m \times n$ matrix is in ***row-reduced echelon form*** if and only if it satisfies the following conditions:

1) In scanning any row from left to right, the first non-zero element encountered (*if any exist*) must be a 1. Call each such element a ***pivot*** (*If it is not 1, remedy by using elementary row operation type 2.*)

2) For any column in which a pivot occurs, all other elements in that column must be zero. (*If they are not, remedy by using elementary row operation type 3.*)

3) Suppose there are r non-zero rows, (*each such row has at least one non-zero element*), $1 \leq r \leq m$. The $m-r$ zero rows must be situated below the r non-zero rows. (*If they are not, remedy by using elementary row operation type 1.*)

4) The r non-zero rows must be arranged in an order such that the pivot in the kth row must lie in a column to the right of the column in which the pivots in rows 1 through $k-1$ lie. More concisely, each pivot must lie to the right of all pivots above it. (*If not, remedy by using elementary row operation type 1.*) .

Example 2.3 The following matrices are in row-reduced echelon form:

$$\begin{pmatrix} 1 & 0 & 2 \\ 0 & 1 & 3 \end{pmatrix} \quad \begin{pmatrix} 1 & 2 & 0 \\ 0 & 0 & 1 \end{pmatrix} \quad \begin{pmatrix} 0 & 1 & 0 \\ 0 & 0 & 1 \end{pmatrix} \quad \begin{pmatrix} 0 & 1 & 2 \\ 0 & 0 & 0 \end{pmatrix} \quad \begin{pmatrix} 0 & 0 & 1 \\ 0 & 0 & 0 \end{pmatrix}$$

$$\begin{pmatrix} 1 & 0 & 0 \\ 0 & 1 & 0 \end{pmatrix} \quad \begin{pmatrix} 0 & 0 & 1 \\ 0 & 0 & 0 \\ 0 & 0 & 0 \end{pmatrix} \quad \begin{pmatrix} 0 & 1 & 2 & 0 & -1 & 1 & 0 & 3 \\ 0 & 0 & 0 & 1 & 0 & 2 & 0 & -2 \\ 0 & 0 & 0 & 0 & 0 & 0 & 1 & -1 \\ 0 & 0 & 0 & 0 & 0 & 0 & 0 & 0 \end{pmatrix} .$$

Example 2.4 The following matrices are ***not*** in row-reduced echelon form:

$$\begin{pmatrix} 1 & 0 \\ 1 & 0 \end{pmatrix} \quad \begin{pmatrix} 2 & 1 \\ 0 & 0 \end{pmatrix} \quad \begin{pmatrix} 0 & 0 & 0 \\ 1 & 0 & 0 \end{pmatrix} \quad \begin{pmatrix} 0 & 0 & 1 & 0 & 2 \\ 0 & 1 & 0 & 2 & -1 \\ 1 & 0 & 0 & 1 & 2 \end{pmatrix} .$$

Each matrix is row equivalent to one and only one row-reduced echelon form matrix. The process of producing the row-reduced echelon form matrix requires the application of a sequence of elementary row operations to the original matrix. Many different sequences will achieve the same goal. Working through the following examples will help to teach you how to achieve the row-reduced echelon forms.

Example 2.5 Find the row-reduced echelon form of each of the matrices in Example 2.4. Note that a number above a $\sim$ indicates which type of elementary row operation has been used in that step.

Solution.

$$\begin{pmatrix} 1 & 0 \\ 1 & 0 \end{pmatrix} \overset{3}{\sim} \begin{pmatrix} 1 & 0 \\ 0 & 0 \end{pmatrix}$$

$$\begin{pmatrix} 2 & 1 \\ 0 & 0 \end{pmatrix} \overset{2}{\sim} \begin{pmatrix} 1 & \frac{1}{2} \\ 0 & 0 \end{pmatrix}$$

$$\begin{pmatrix} 0 & 0 & 0 \\ 1 & 0 & 0 \end{pmatrix} \overset{1}{\sim} \begin{pmatrix} 1 & 0 & 0 \\ 0 & 0 & 0 \end{pmatrix}$$

$$\begin{pmatrix} 0 & 0 & 1 & 0 & 2 \\ 0 & 1 & 0 & 2 & -1 \\ 1 & 0 & 0 & 1 & 2 \end{pmatrix} \overset{1}{\sim} \begin{pmatrix} 1 & 0 & 0 & 1 & 2 \\ 0 & 1 & 0 & 2 & -1 \\ 0 & 0 & 1 & 0 & 2 \end{pmatrix} .$$

Example 2.6 Find the row-reduced echelon form of

$$\begin{pmatrix} 3 & 2 \\ -2 & 1 \end{pmatrix}.$$

Solution.

$$\begin{pmatrix} 3 & 2 \\ -2 & 1 \end{pmatrix} \overset{2}{\sim} \begin{pmatrix} 1 & \frac{2}{3} \\ -2 & 1 \end{pmatrix} \overset{3}{\sim} \begin{pmatrix} 1 & \frac{2}{3} \\ 0 & \frac{7}{3} \end{pmatrix} \overset{2}{\sim} \begin{pmatrix} 1 & \frac{2}{3} \\ 0 & 1 \end{pmatrix} \overset{3}{\sim} \begin{pmatrix} 1 & 0 \\ 0 & 1 \end{pmatrix}.$$

Example 2.7 Find the row-reduced echelon form of

$$\begin{pmatrix} 2 & 1 & 0 \\ 1 & 0 & -1 \\ 0 & 1 & -2 \\ -1 & 1 & 2 \end{pmatrix}.$$

Solution.

$$\begin{pmatrix} 2 & 1 & 0 \\ 1 & 0 & -1 \\ 0 & 1 & -2 \\ -1 & 1 & 2 \end{pmatrix} \overset{1}{\sim} \begin{pmatrix} 1 & 0 & -1 \\ 2 & 1 & 0 \\ 0 & 1 & -2 \\ -1 & 1 & 2 \end{pmatrix} \overset{3}{\sim} \begin{pmatrix} 1 & 0 & -1 \\ 0 & 1 & 2 \\ 0 & 1 & -2 \\ 0 & 1 & 1 \end{pmatrix} \overset{3}{\sim}$$

$$\begin{pmatrix} 1 & 0 & -1 \\ 0 & 1 & 2 \\ 0 & 0 & -4 \\ 0 & 0 & -1 \end{pmatrix} \overset{2}{\sim} \begin{pmatrix} 1 & 0 & -1 \\ 0 & 1 & 2 \\ 0 & 0 & 1 \\ 0 & 0 & -1 \end{pmatrix} \overset{3}{\sim} \begin{pmatrix} 1 & 0 & 0 \\ 0 & 1 & 0 \\ 0 & 0 & 1 \\ 0 & 0 & 0 \end{pmatrix}.$$

Example 2.8 Find the row-reduced echelon form of

$$\begin{pmatrix} 0 & 1 & -1 & 2 & 0 \\ 0 & -1 & 1 & -1 & -1 \\ 0 & 2 & -2 & 7 & -2 \\ 0 & -2 & 2 & -5 & 3 \end{pmatrix}.$$

Solution.

$$\begin{pmatrix} 0 & 1 & -1 & 2 & 0 \\ 0 & -1 & 1 & -1 & -1 \\ 0 & 2 & -2 & 7 & -2 \\ 0 & -2 & 2 & -5 & 3 \end{pmatrix} \overset{3}{\sim} \begin{pmatrix} 0 & 1 & -1 & 2 & 0 \\ 0 & 0 & 0 & 1 & -1 \\ 0 & 0 & 0 & 3 & -2 \\ 0 & 0 & 0 & -1 & 3 \end{pmatrix} \overset{3}{\sim}$$

$$\begin{pmatrix} 0 & 1 & -1 & 0 & 2 \\ 0 & 0 & 0 & 1 & -1 \\ 0 & 0 & 0 & 0 & 1 \\ 0 & 0 & 0 & 0 & 2 \end{pmatrix} \overset{3}{\sim} \begin{pmatrix} 0 & 1 & -1 & 0 & 0 \\ 0 & 0 & 0 & 1 & 0 \\ 0 & 0 & 0 & 0 & 1 \\ 0 & 0 & 0 & 0 & 0 \end{pmatrix}.$$

2.6 Exercises

1. Find the coefficient matrix and augmented matrix of each of the following systems of equations

a) $2x_1 + 3x_2 + 5x_3 + 7x_4 = 8$

b)
$$\begin{aligned} 3x + 2y &= 1 \\ 4x + 3y &= 2 \\ 5x + 4y &= 3 \\ 6x + 5y &= 4 \end{aligned}$$

c)
$$\begin{aligned} 2x + 2y - z &= -1 \\ x + y + 2z &= 2 \\ x + y + z &= 1 \end{aligned}$$

d)
$$\begin{aligned} 2x_1 + x_2 - 4x_3 + 4x_4 &= 1 \\ x_1 + 2x_2 + x_3 + 4x_4 &= 2 \end{aligned}$$

e)
$$\begin{aligned} x + 2y &= 0 \\ x + 3y &= 0 \\ 2x + 5y &= 0 \end{aligned}$$

f)
$$\begin{aligned} 3u + 5v &= 4 \\ 2u + 7v &= 1 \end{aligned}.$$

2. Use Gaussian Elimination on A to find the row-reduced echelon form matrix equivalent to it if A is given by

a) $\begin{pmatrix} 1 & 2 & 1 & 1 \\ 1 & 3 & 2 & 4 \end{pmatrix}$ **b)** $\begin{pmatrix} 1 & 1 & 1 & 1 \\ 2 & 3 & 3 & 3 \\ 3 & 4 & 2 & 2 \end{pmatrix}$ **c)** $\begin{pmatrix} 2 & 1 & 2 \\ -2 & 2 & 1 \\ 1 & 2 & -2 \end{pmatrix}$

d) $\begin{pmatrix} 2 & 2 & -1 & -1 \\ 1 & 1 & 2 & 2 \\ 1 & 1 & 1 & 1 \end{pmatrix}$ **e)** $\begin{pmatrix} 2 & 2 & -1 & 1 \\ 1 & 1 & 2 & 2 \\ 1 & 1 & 1 & 1 \end{pmatrix}$ **f)** $\begin{pmatrix} 0 & 2 & -1 & 1 \\ 0 & 1 & 1 & 5 \end{pmatrix}$.

3. Assume each of the following matrices is an augmented matrix. Write the system of equations represented by each matrix.

a) $\left(\begin{array}{cc|c} 1 & 0 & 0 \\ 0 & 1 & 1 \end{array}\right)$ **b)** $\left(\begin{array}{c|c} 1 & 1 \\ 2 & 0 \end{array}\right)$ **c)** $\left(\begin{array}{ccc|c} 1 & 2 & -1 & 1 \\ 3 & -1 & 0 & 2 \\ 0 & 1 & 1 & -1 \end{array}\right)$ **d)** $\left(\begin{array}{cc|c} 1 & -1 & 2 \\ 2 & 0 & -1 \\ 1 & 1 & 0 \end{array}\right)$.

4. Starting on page 19, four conditions are given which a matrix must satisfy to be in row-reduced echelon form. For each matrix in Example 2.4, page 20, state which condition fails to be satisfied.

5. Determine whether or not the following two systems are equivalent:

$$\begin{aligned} x_1 - x_3 - x_4 &= 0 \\ -x_1 + x_2 + x_4 &= 1 \\ x_3 - x_4 &= 1 \\ x_1 + x_2 + x_4 &= -1 \end{aligned}$$

and

$$\begin{aligned} x_1 - 2x_4 &= 1 \\ 2x_1 + x_2 - x_3 &= -1 \\ x_1 - x_3 - x_4 &= 0 \\ -x_1 + x_2 + x_3 &= 2 \end{aligned} .$$

6. Perform row operations on the matrix $A = \begin{pmatrix} 0 & 1 & 1 & 5 \\ 0 & 1 & 2 & 7 \\ 0 & 1 & 2 & 7 \end{pmatrix}$ to put it into row-reduced echelon form.

7. The matrix

$$A = \begin{pmatrix} 0 & 0 & a_{13} & a_{14} & a_{15} \\ 0 & 0 & a_{23} & a_{24} & a_{25} \\ 0 & 0 & a_{33} & a_{34} & a_{35} \end{pmatrix}$$

can be row reduced to a matrix R which is in ***row reduced echelon form***. Find R if you know that it has ***three*** non-zero rows.

3 Varieties of Systems of Linear Equations

Outline of Key Ideas

Consistent versus inconsistent systems:

inconsistent systems have no solutions

consistent systems have solutions.

Solutions of consistent systems:

the single solution case: exactly one solution

the underdetermined case: infinitely many solutions.

How to solve systems of linear equations: a summary.

Varieties of Systems of Linear Equations

We have seen that a general procedure for solving the system of linear equations

$$A\mathbf{x} = \mathbf{b}$$

is to apply a sequence of elementary row operations to the corresponding augmented matrix $[A|\mathbf{b}]$ to obtain its row-reduced echelon form $[A_0|\mathbf{b}_0]$. This form corresponds to the equivalent system of equations

$$A_0\mathbf{x} = \mathbf{b}_0$$

that is simplest to solve. Its solution set is also the solution set of the original system since the two systems are equivalent. What possible situations arise in such a procedure?

3.1 Consistent Versus Inconsistent Systems

At some point during the procedure of obtaining the row-reduced echelon form, one of the rows of the augmented matrix may become

$$\left(\begin{array}{cccc|c} 0 & 0 & \cdots & 0 & a \end{array} \right) \quad . \tag{3.1}$$

The corresponding equation of the system is

$$0x_1 + 0x_2 + \cdots + 0x_n = a \quad . \tag{3.2}$$

If $a \neq 0$, there is no set of values for the unknowns $\{x_1, x_2, \ldots, x_n\}$ which satisfies this equation. Consequently, there is ***no*** solution to the system of equations. The system is said to be ***inconsistent***, a term indicating that the system makes a requirement (Equation (3.2)) which no vector $\mathbf{x}$ can satisfy. Consider the following examples.

Example 3.1 The system with one equation and one unknown

$$0x_1 = 1 \quad .$$

is an inconsistent system. There is no value for x_1 which satisfies the equation.

Example 3.2 The system

$$\begin{aligned} 3x_1 + 2x_2 &= 1 \\ 3x_1 + 2x_2 &= -1 \end{aligned} .$$

is clearly inconsistent. It demands that the same quantity be both 1 and -1. The row reduction procedure gives us:

$$\left(\begin{array}{cc|c} 3 & 2 & 1 \\ 3 & 2 & -1 \end{array}\right) \sim \left(\begin{array}{cc|c} 3 & 2 & 1 \\ 0 & 0 & -2 \end{array}\right) .$$

The second row is the type shown in (3.1) and represents an equation of the type shown in (3.2) which is impossible to satisfy. This confirms that the system in inconsistent.

Example 3.3 It is not immediately obvious that the system

$$\begin{aligned} x_1 \qquad\quad - x_3 &= 2 \\ x_1 + x_2 \qquad\quad &= -1 \\ 3x_1 + 2x_2 - x_3 &= 1 \end{aligned} .$$

is inconsistent. The row reduction process gives us

$$\left(\begin{array}{ccc|c} 1 & 0 & -1 & 2 \\ 1 & 1 & 0 & -1 \\ 3 & 2 & -1 & 1 \end{array}\right) \sim \left(\begin{array}{ccc|c} 1 & 0 & -1 & 2 \\ 0 & 1 & 1 & -3 \\ 0 & 2 & 2 & -5 \end{array}\right) \sim \left(\begin{array}{ccc|c} 1 & 0 & -1 & 2 \\ 0 & 1 & 1 & -3 \\ 0 & 0 & 0 & 1 \end{array}\right) .$$

The third row in the last matrix shows that the system is inconsistent.

The row reduction process clearly determines the inconsistency of a system. We obtain the immediate conclusion that the system has ***no*** solutions (*the solution set is empty*).

If either no row of the type (3.1) is encountered in the row reduction process or for every such row encountered we have $a = 0$, then the system is called ***consistent***. In the case $a = 0$, the equation corresponding to that row plays no part in determining the solution set of the system. Recall that such a zero row will eventually be placed out of the way at the bottom of the row-reduced echelon form of the augmented matrix.

Example 3.4 The one-equation and one-unknown system

$$0x_1 = 0$$

has the augmented matrix

$$\left(\begin{array}{c|c} 0 & 0 \end{array}\right) .$$

The system is consistent. The only equation places no demand on the solution. (*What is the solution set for this system?*)

Example 3.5

$$\left(\begin{array}{rrr|r} 1 & 0 & -1 & 2 \\ 1 & 1 & 0 & -1 \\ 3 & 2 & -1 & 0 \\ -2 & -1 & 1 & -1 \end{array}\right) \sim \left(\begin{array}{rrr|r} 1 & 0 & -1 & 2 \\ 0 & 1 & 1 & -3 \\ 0 & 2 & 2 & -6 \\ 0 & -1 & -1 & 3 \end{array}\right) \sim \left(\begin{array}{rrr|r} 1 & 0 & -1 & 2 \\ 0 & 1 & 1 & -3 \\ 0 & 0 & 0 & 0 \\ 0 & 0 & 0 & 0 \end{array}\right)$$

so the corresponding system is consistent.

A special class of systems occurs frequently enough to deserve some individual attention. Each member of this class is called a ***homogeneous system of equations*** and has the form

$$A\mathbf{x} = \mathbf{0} \quad . \tag{3.3}$$

Its augmented matrix is $[A|\mathbf{0}]$, i.e. the last column is filled with zeros. If you consider the effect on that last column of each of the three types of elementary row operations, your conclusion will be that the last column never changes from zero! So the inconsistency condition is never encountered. ***A homogeneous system is always consistent!***

We have seen that inconsistent systems have no solutions. We shall see that ***homogeneous systems always have at least one solution.*** (*For example, name one vector that is in the solution set of every homogeneous system*). So determination of the consistency or inconsistency of a system is clearly an important step in seeking the solution set of a system.

3.2 Solving the Consistent System

We have stated that a consistent system has at least one solution. It may have ***exactly one*** solution or it may have ***infinitely many*** solutions. Let us see how these results can occur.

First disregard any zero rows from the row-reduced echelon form of the augmented matrix. These zero rows correspond to equations in the original system which were eliminated by a sequence of elementary row operations. To illustrate this statement, suppose the second and fourth equations of a system are given by

$$\begin{array}{ll} 2^{\text{nd}}: & 3x_1 + x_2 - 2x_3 - x_4 = 2 \\ 4^{\text{th}}: & 6x_1 + 2x_2 - 4x_3 - 2x_4 = 4 \end{array} . \tag{3.4}$$

The 4th equation is clearly just twice the 2nd equation and can be eliminated by subtracting twice the 2nd equation. Then the corresponding procedure applied to the augmented matrix will produce the following:

$$\left(\begin{array}{cccc|c} \cdot & \cdot & \cdot & \cdot & \cdot \\ 3 & 1 & -2 & -1 & 2 \\ \cdot & \cdot & \cdot & \cdot & \cdot \\ 6 & 2 & -4 & -2 & 4 \\ \cdot & \cdot & \cdot & \cdot & \cdot \\ \cdot & \cdot & \cdot & \cdot & \cdot \end{array}\right) \sim \left(\begin{array}{cccc|c} \cdot & \cdot & \cdot & \cdot & \cdot \\ 3 & 1 & -2 & -1 & 2 \\ \cdot & \cdot & \cdot & \cdot & \cdot \\ 0 & 0 & 0 & 0 & 0 \\ \cdot & \cdot & \cdot & \cdot & \cdot \\ \cdot & \cdot & \cdot & \cdot & \cdot \end{array}\right) .$$

Thus for a consistent system, the number r of non-zero rows in the row- reduced echelon form of the coefficient matrix corresponds to the number equations in the system after all the elementary operations have been completed. Since every matrix is equivalent to one and only one row reduced echelon form it follows that every linear system of equations is equivalent to a unique set of r equations. All the equations in the original system are ***linear combinations*** of these r equations. For instance the system

$$\begin{aligned} x_1 + 0x_2 - x_3 &= 2 \\ x_1 + x_2 + 0x_3 &= -1 \\ 3x_1 + 2x_2 - x_3 &= 0 \\ -2x_1 - x_2 + x_3 &= -1 \end{aligned} \tag{3.5}$$

corresponding to the augmented matrix in Example 3.5 is equivalent to

$$\begin{aligned} x_1 + 0x_2 - x_3 &= 2 \\ x_2 + x_3 &= -3 \end{aligned} . \tag{3.6}$$

Then Equation (3.5) can be given by the following

$$\begin{aligned} 1(x_1 + 0x_2 - x_3) + 0(x_2 + x_3) &= 1(2) + 0(-3) \\ 1(x_1 + 0x_2 - x_3) + 1(x_2 + x_3) &= 1(2) + 1(-3) \\ 3(x_1 + 0x_2 - x_3) + 2(x_2 + x_3) &= 3(2) + 2(-3) \\ -2(x_1 + 0x_2 - x_3) - 1(x_2 + x_3) &= -2(2) - 1(-3) \end{aligned} .$$

Thus, the number $r = 2$ of non-zero rows in the matrix

$$\left(\begin{array}{ccc|c} 1 & 0 & -1 & 2 \\ 0 & 1 & 1 & -3 \end{array}\right)$$

corresponds to the $r = 2$ equations in (3.6) which are equivalent to (3.5).

The number r is called the ***rank*** of the coefficient matrix A of the system $A\mathbf{x} = \mathbf{b}$. Comparison of the rank r to the number of unknowns n of the system determines the solution situation.

3.3 The Single Solution Case

When $r = n$, the system of equations corresponding to the row-reduced echelon form of the augmented matrix (*with zero rows ignored*) has the form

(3.7)
$$\begin{aligned} x_1 &= b'_1 \\ x_2 &= b'_2 \\ x_3 &= b'_3 \\ &\vdots \\ x_n &= b'_n \ . \end{aligned}$$

No equation involves more than one unknown and each unknown is involved in one equation. Clearly, in this case, there is exactly one solution, namely (*written in terms of column vectors*)

(3.8)
$$\mathbf{x} = \mathbf{b}' \ .$$

Example 3.6 The system of one equation and one unknown given by $3x = 5$ has the augmented matrix $\left(\begin{array}{cc} 3 & 5 \end{array}\right)$

$$\left(\begin{array}{c|c} 3 & 5 \end{array}\right) \sim \left(\begin{array}{c|c} 1 & \frac{5}{3} \end{array}\right)$$

so the solution set $\mathcal{S} = \{\frac{5}{3}\}$.

Example 3.7 Suppose the row reduction of the augmented matrix of a system is given by

$$\left(\begin{array}{rrr|r} 1 & 1 & 2 & -1 \\ -1 & 0 & 2 & -3 \\ 0 & 1 & 2 & -2 \\ 2 & -1 & 1 & 1 \end{array}\right) \sim \left(\begin{array}{rrr|r} 1 & 1 & 2 & -1 \\ 0 & 1 & 4 & -4 \\ 0 & 1 & 2 & -2 \\ 0 & -3 & -3 & 3 \end{array}\right) \sim \left(\begin{array}{rrr|r} 1 & 1 & 2 & -1 \\ 0 & 1 & 4 & -4 \\ 0 & 0 & -2 & 2 \\ 0 & 0 & 9 & -9 \end{array}\right) \sim$$

$$\left(\begin{array}{rrr|r} 1 & 0 & 0 & 1 \\ 0 & 1 & 0 & 0 \\ 0 & 0 & 1 & -1 \\ 0 & 0 & 0 & 0 \end{array}\right).$$

Then the equivalent system of equations is

$$\begin{aligned} x_1 &= 1 \\ x_2 &= 0 \\ x_3 &= -1 \ . \end{aligned}$$

and so the solution set is $\mathcal{S} = \left\{ \begin{pmatrix} 1 \\ 0 \\ -1 \end{pmatrix} \right\}$.

Example 3.8 Suppose the row reduction of the augmented matrix for a homogeneous system $A\mathbf{x} = \mathbf{0}$ is given by

$$\left(\begin{array}{rrr|r} 1 & 2 & 1 & 0 \\ 0 & 1 & -1 & 0 \\ -1 & 0 & 1 & 0 \end{array}\right) \sim \left(\begin{array}{rrr|r} 1 & 2 & 1 & 0 \\ 0 & 1 & -1 & 0 \\ 0 & 2 & 2 & 0 \end{array}\right) \sim \left(\begin{array}{rrr|r} 1 & 0 & 3 & 0 \\ 0 & 1 & -1 & 0 \\ 0 & 0 & 4 & 0 \end{array}\right) \sim$$

$$\left(\begin{array}{rrr|r} 1 & 0 & 3 & 0 \\ 0 & 1 & -1 & 0 \\ 0 & 0 & 1 & 0 \end{array}\right) \sim$$

$$\left(\begin{array}{rrr|r} 1 & 0 & 0 & 0 \\ 0 & 1 & 0 & 0 \\ 0 & 0 & 1 & 0 \end{array}\right).$$

Then the equivalent system of equations is

$$\begin{aligned} x_1 &= 0 \\ x_2 &= 0 \\ x_3 &= 0 \end{aligned}$$

and so the solution set is $\mathcal{S} = \left\{ \begin{pmatrix} 0 \\ 0 \\ 0 \end{pmatrix} \right\}$.

Clearly this will always occur for a homogeneous system in which the number of truly independent equations equals the number of unknowns. It is often described by the statement that such a system has only the trivial solution. The name trivial solution refers to the $\mathbf{x} = \mathbf{0}$ vector which is obviously a solution to any homogeneous system $A\mathbf{x} = \mathbf{0}$. Note that the row reduction procedure used in this example is required to establish that the obvious $\mathbf{x} = \mathbf{0}$ solution is the ***only*** solution.

In practice it is not necessary to row reduce the augmented matrix all the way to the echelon form. In fact, it is computationally efficient to perform the same row reduction process ***except*** to not seek to make zero all the column elements ***above*** a pivot. This will ***not*** produce the theoretically simplest equivalent system of equations, but it will produce a system which can be solved by the method of back-substitution.

We will repeat the solution of the system in Example 3.7 to illustrate this idea:

Example 3.9

$$\left(\begin{array}{rrr|r} 1 & 1 & 2 & -1 \\ -1 & 0 & 2 & -3 \\ 0 & 1 & 2 & -2 \\ 2 & -1 & 1 & 1 \end{array}\right) \sim \left(\begin{array}{rrr|r} 1 & 1 & 2 & -1 \\ 0 & 1 & 4 & -4 \\ 0 & 1 & 2 & -2 \\ 0 & -3 & -3 & 3 \end{array}\right) \sim \left(\begin{array}{rrr|r} 1 & 1 & 2 & -1 \\ 0 & 1 & 4 & -4 \\ 0 & 0 & -2 & 2 \\ 0 & 0 & 9 & -9 \end{array}\right) \sim$$

$$\left(\begin{array}{rrr|r} 1 & 1 & 2 & -1 \\ 0 & 1 & 4 & -4 \\ 0 & 0 & 1 & -1 \\ 0 & 0 & 0 & 0 \end{array}\right) .$$

At this stage, all elements below the pivots are zero. The system of equations is now

$$\begin{aligned} x_1 + x_2 + 2x_3 &= -1 \\ x_2 + 4x_3 &= -4 \\ x_3 &= -1 \quad . \end{aligned}$$

The idea of solving by back-substitution now follows from the special form of the system. The ***last*** equation is already solved for x_3. Substituting this value for x_3 into the second equation gives us

$$x_2 + 4(-1) = x_2 - 4 = -4, \quad \text{so} \quad x_2 = 0 \quad .$$

Now using the values obtained for x_2 and x_3 in the first equation, we have

$$x_1 + 0 + 2(-1) = -1, \quad \text{so} \quad x_1 = 1 \quad .$$

Then the solution set is $\mathcal{S} = \left\{ \begin{pmatrix} 1 \\ 0 \\ -1 \end{pmatrix} \right\}$ as in Example 3.7.

Observe that if we start with the last non-zero row of

$$\left(\begin{array}{ccc|c} 1 & 1 & 2 & -1 \\ 0 & 1 & 4 & -4 \\ 0 & 0 & 1 & -1 \\ 0 & 0 & 0 & 0 \end{array} \right)$$

and work from bottom to top to put it into row-reduced echelon form we obtain

$$\left(\begin{array}{ccc|c} 1 & 1 & 2 & -1 \\ 0 & 1 & 4 & -4 \\ 0 & 0 & 1 & -1 \\ 0 & 0 & 0 & 0 \end{array} \right) \sim \left(\begin{array}{ccc|c} 1 & 1 & 0 & 1 \\ 0 & 1 & 0 & 0 \\ 0 & 0 & 1 & -1 \\ 0 & 0 & 0 & 0 \end{array} \right) \sim \left(\begin{array}{ccc|c} 1 & 0 & 0 & 1 \\ 0 & 1 & 0 & 0 \\ 0 & 0 & 1 & -1 \\ 0 & 0 & 0 & 0 \end{array} \right)$$

as before. These operations are equivalent to back substitution.

As a final check, it can be confirmed that

$$\begin{pmatrix} 1 & 1 & 2 \\ -1 & 0 & 2 \\ 0 & 1 & 2 \\ 2 & -1 & 1 \end{pmatrix} \begin{pmatrix} 1 \\ 0 \\ -1 \end{pmatrix} = \begin{pmatrix} -1 \\ -3 \\ -2 \\ 1 \end{pmatrix} \quad .$$

3.4 The Underdetermined Case

When $r < n$, the number of truly independent equations is less than the number of unknowns. Such a system is called ***underdetermined***. We will use a specific example to discuss the characterization of the solution set of the underdetermined system.

Suppose that (*ignoring zero rows*) the row-reduced echelon form of the augmented matrix of the system is

$$\left(\begin{array}{cccccc|c} 1 & -2 & 0 & 0 & 2 & 0 & 4 \\ 0 & 0 & 1 & 0 & -1 & 0 & 2 \\ 0 & 0 & 0 & 1 & 1 & 0 & 0 \\ 0 & 0 & 0 & 0 & 0 & 1 & -1 \end{array}\right) \quad . \tag{3.9}$$

In the language of equations this is equivalent to

$$\begin{aligned} x_1 - 2x_2 + 2x_5 &= 4 \\ x_3 - x_5 &= 2 \\ x_4 + x_5 &= 0 \\ x_6 &= -1 \end{aligned} \quad . \tag{3.10}$$

So our reduced system has four independent equations ($r = 4$) and six unknowns ($n = 6$). What are its solution vectors? To find them, note that the four unknowns x_1, x_3, x_4, x_6, corresponding to the four pivot positions in the augmented matrix (3.9) each occur in one and only one of the four equations (3.10). (*This is guaranteed by the requirement that all other elements in the column of a pivot be zero.*) As a consequence, one can solve for each one of the four variables in its own equation. The system then has the form

$$\begin{aligned} x_1 &= 4 + 2x_2 - 2x_5 \\ x_3 &= 2 + x_5 \\ x_4 &= -x_5 \\ x_6 &= -1 \quad . \end{aligned} \tag{3.11}$$

This form makes very clear that a solution vector $\mathbf{x}$ for (3.11) (*and hence the original system*) will be obtained by choosing arbitrary values for the second and fifth components (x_2 and x_5) and then the values of the remaining components (x_1, x_3, x_4, and x_6) are calculated by the equations (3.11). For example, choose $x_2 = 0$ and $x_5 = 0$, then $\mathbf{x} = \langle 4, 0, 2, 0, 0, -1\rangle$ is a solution vector. Now choose $x_2 = 2$ and $x_5 = -1$, then the corresponding solution vector is $\mathbf{x} = \langle 10, 2, 1, 1, -1, -1\rangle$. How many solution vectors are there? The infinity of choices for values of x_2 and x_5 produce ***an infinite number of solution vectors.***

Some terminology for the underdetermined system situation is useful. The unknowns associated with the pivot positions in the row-reduced echelon form of the augmented matrix are called ***basic variables.*** In our illustrative example, $\{x_1, x_3, x_4, x_6\}$ is the set of basic variables. The remaining unknowns are called ***free variables*** relating to the freedom of choice in their values. In our example, x_2 and x_5 are the free variables.

The situation involved in the underdetermined system having n unknowns and rank r may be stated as follows. There are r basic variables and $n-r$ free variables. Each basic variable occurs in one and only one equation of the system corresponding to the row-reduced echelon form of the augmented matrix. Solve each of the r equations for its one basic variable. As a consequence, each basic variable is expressed entirely in terms of constants and the free variables. Then each possible solution vector of the system can be obtained by choosing a real number value for each free variable vector component and calculating the resulting values for the basic variable components. There are clearly infinitely many solution vectors corresponding to the infinite choice of values for the free variables.

The solution set has infinitely many members. Each member is an $n-$component vector, but not every $n-$component vector is a member. How shall we conveniently represent this set? Using our example for illustration, we can write the solution set as follows after setting $x_2 = \alpha$ and $x_3 = \beta$:

$$\mathcal{S} = \left\{ \left(\begin{array}{c} 4+2\alpha-2\beta \\ \alpha \\ 2+\beta \\ -\beta \\ \beta \\ -1 \end{array} \right) \middle| \; \alpha, \beta \in \mathcal{R} \right\} . \tag{3.12}$$

This notation is interpreted as follows:

$\mathcal{S}$ is the set of all possible 6×1 column vectors with elements expressed as shown in terms of two free variables x_2 and x_5, whose values are arbitrarily chosen real numbers. $x_2 = \alpha$ and $x_5 = \beta$.

An alternative representation of the typical solution vector has a useful interpretation. Using the properties of matrix algebra, the solution column vector can be written

$$\left(\begin{array}{c} 4+2\alpha-2\beta \\ \alpha \\ 2+\beta \\ -\beta \\ \beta \\ -1 \end{array} \right) = \left(\begin{array}{c} 4 \\ 0 \\ 2 \\ 0 \\ 0 \\ -1 \end{array} \right) + \left(\begin{array}{c} 2\alpha \\ \alpha \\ 0 \\ 0 \\ 0 \\ 0 \end{array} \right) + \left(\begin{array}{c} -2\beta \\ 0 \\ \beta \\ -\beta \\ \beta \\ 0 \end{array} \right) \tag{3.13}$$

$$(3.14) \qquad = \begin{pmatrix} 4 \\ 0 \\ 2 \\ 0 \\ 0 \\ -1 \end{pmatrix} + \alpha \begin{pmatrix} 2 \\ 1 \\ 0 \\ 0 \\ 0 \\ 0 \end{pmatrix} + \beta \begin{pmatrix} -2 \\ 0 \\ 1 \\ -1 \\ 1 \\ 0 \end{pmatrix} .$$

Define the 6−component vectors **a**, **b**, and **c** as

$$(3.15) \qquad \begin{aligned} \mathbf{a} &= \langle 4,0,2,0,0,-1\rangle^T \\ \mathbf{b} &= \langle 2,1,0,0,0,0\rangle^T \\ \mathbf{c} &= \langle -2,0,1,-1,1,0\rangle^T \end{aligned}$$

then (3.13) may be written in vector notation as

$$(3.16) \qquad \mathbf{x} = \mathbf{a} + \alpha\mathbf{b} + \beta\mathbf{c}$$

This characterizes each solution column vector **x** in terms of three specific constant column vectors **a**, **b**, and **c**. The vector **x** is said to be a ***linear combination*** of the vectors **a**, **b**, and **c** and the solution set $\mathcal{S}$ is given by

$$(3.17) \qquad \mathcal{S} = \{\mathbf{a} + \alpha\mathbf{b} + \beta\mathbf{c} \mid \alpha, \beta \in \mathcal{R}\} \quad .$$

Personal preference or the application will determine whether the form of (3.12) or (3.17) is used to describe the solution set of an underdetermined system.

Example 3.10 Consider the system of one equation with two unknowns:

$$x_1 + x_2 = 2 \quad .$$

Its augmented matrix $\left(1 \quad 1 \,|\, 2 \right)$ is already in row-reduced echelon form with the $(1,1)$ element as the pivot. So x_1 is the basic variable and x_2 is the free variable. So we set $x_2 = \alpha$ and solve for x_1:

$$x_1 = 2 - x_2 = 2 - \alpha \quad .$$

Consequently, the solution set can be written as

$$\mathcal{S} = \left\{ \begin{pmatrix} 2-\alpha \\ \alpha \end{pmatrix} \middle| \, \alpha \in \mathcal{R} \right\} \quad .$$

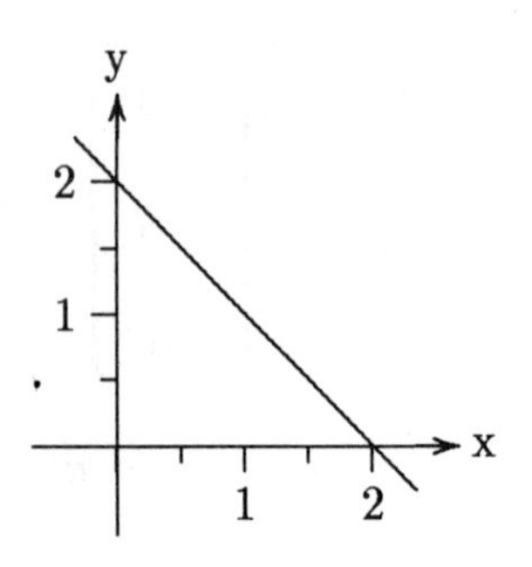

Figure 3.1

For example, $\begin{pmatrix} 2 \\ 0 \end{pmatrix}$ and $\begin{pmatrix} 1.9 \\ 0.1 \end{pmatrix}$ are solutions to the system. (*Give 3 more solutions.*) In fact, the solutions of this system are the vectors in $\mathcal{R}^2$ whose components add up to 2. If each solution vector of the system is used as a position vector of a point in the plane (*i.e. place the solution vector beginning at the origin, then the tip defines a point in the plane*), the locus of points in the plane described by the solution set will be a straight line. Writing

$$\begin{pmatrix} 2-\alpha \\ \alpha \end{pmatrix} = \begin{pmatrix} 2 \\ 0 \end{pmatrix} + \alpha \begin{pmatrix} -1 \\ 1 \end{pmatrix}$$

we have a vector equation for the straight line through the point $\langle 2, 0\rangle^T$ with a direction vector $\langle -1, 1\rangle^T$. If the horizontal axis is labeled x and the vertical one y then upon elimination of the parameter α we find that the **Cartesian equation** of the line is given by $y = -x + 2$. (*See Figure 3.1*)

Example 3.11 Suppose a system has seven equations and six unknowns. Is it inconsistent, underdetermined, or is its rank six? It is usually necessary to perform row reduction to find out. Suppose its augmented matrix is

$$\left(\begin{array}{rrrrrr|r} 2 & 2 & -2 & -1 & 1 & 2 & 5 \\ -1 & -1 & 1 & 1 & 1 & -2 & 0 \\ 1 & 1 & -1 & 2 & -1 & 5 & 6 \\ -2 & -2 & 2 & 1 & 0 & -3 & -4 \\ -1 & -1 & 1 & 3 & -2 & 3 & 1 \\ -3 & -3 & 3 & 2 & 1 & -5 & -4 \\ -2 & -2 & 2 & 4 & 0 & 0 & 2 \end{array}\right).$$

Choosing the $(3, 1)$ element as our first column pivot, interchange rows 1 and 3 and use the pivot to zero the remainder of column 1 to get

$$\left(\begin{array}{cccccc|c} 1 & 1 & -1 & 2 & -1 & 5 & 6 \\ 0 & 0 & 0 & 3 & 0 & 3 & 6 \\ 0 & 0 & 0 & -5 & 3 & -8 & -7 \\ 0 & 0 & 0 & 5 & -2 & 7 & 8 \\ 0 & 0 & 0 & 5 & -3 & 8 & 7 \\ 0 & 0 & 0 & 8 & -2 & 10 & 14 \\ 0 & 0 & 0 & 8 & -2 & 10 & 14 \end{array}\right).$$

Choose the $(2,4)$ element as the next pivot, multiply row 2 by $1/3$ and use the pivot to get

$$\left(\begin{array}{cccccc|c} 1 & 1 & -1 & 2 & -1 & 5 & 6 \\ 0 & 0 & 0 & 1 & 0 & 1 & 2 \\ 0 & 0 & 0 & 0 & 3 & -3 & 3 \\ 0 & 0 & 0 & 0 & -2 & 2 & -2 \\ 0 & 0 & 0 & 0 & -3 & 3 & -3 \\ 0 & 0 & 0 & 0 & -2 & 2 & -2 \\ 0 & 0 & 0 & 0 & -2 & 2 & -2 \end{array}\right).$$

Choose the $(3,5)$ element as the next pivot, multiply row 3 by $1/3$ and use the pivot to get

$$\left(\begin{array}{cccccc|c} 1 & 1 & -1 & 2 & -1 & 5 & 6 \\ 0 & 0 & 0 & 1 & 0 & 1 & 2 \\ 0 & 0 & 0 & 0 & 1 & -1 & 1 \\ 0 & 0 & 0 & 0 & 0 & 0 & 0 \\ 0 & 0 & 0 & 0 & 0 & 0 & 0 \\ 0 & 0 & 0 & 0 & 0 & 0 & 0 \\ 0 & 0 & 0 & 0 & 0 & 0 & 0 \end{array}\right).$$

Finally using the pivots in reverse order to eliminate non-zero column elements above them, we get the row-reduced echelon form

$$\left(\begin{array}{cccccc|c} 1 & 1 & -1 & 0 & 0 & 2 & 3 \\ 0 & 0 & 0 & 1 & 0 & 1 & 2 \\ 0 & 0 & 0 & 0 & 1 & -1 & 1 \\ 0 & 0 & 0 & 0 & 0 & 0 & 0 \\ 0 & 0 & 0 & 0 & 0 & 0 & 0 \\ 0 & 0 & 0 & 0 & 0 & 0 & 0 \\ 0 & 0 & 0 & 0 & 0 & 0 & 0 \end{array}\right).$$

This corresponds to an equivalent system of three equations in six unknowns, so we have the consistent underdetermined system

$$\begin{aligned} x_1 + x_2 - x_3 + 2x_6 &= 3 \\ x_4 + x_6 &= 2 \\ x_5 - x_6 &= 1 \ . \end{aligned}$$

Solve for the basic variables x_1, x_4, x_5 to obtain

$$\begin{aligned} x_1 &= 3 - x_2 + x_3 - 2x_6 \\ x_4 &= 2 - x_6 \\ x_5 &= 1 + x_6 \end{aligned}$$

and set the free variables $x_2 = \alpha$, $x_3 = \beta$, and $x_6 = \gamma$. Then the solution set is given by

$$\mathcal{S} = \left\{ \left(\begin{array}{c} 3 - \alpha + \beta - 2\gamma \\ \alpha \\ \beta \\ 2 - \gamma \\ 1 + \gamma \\ \gamma \end{array} \right) \middle| \ \alpha, \beta, \gamma \in \mathcal{R} \right\} \ .$$

The typical solution vector may also be written

$$\mathbf{x} = \left(\begin{array}{c} 3 \\ 0 \\ 0 \\ 2 \\ 1 \\ 0 \end{array} \right) + \alpha \left(\begin{array}{c} -1 \\ 1 \\ 0 \\ 0 \\ 0 \\ 0 \end{array} \right) + \beta \left(\begin{array}{c} 1 \\ 0 \\ 1 \\ 0 \\ 0 \\ 0 \end{array} \right) + \gamma \left(\begin{array}{c} -2 \\ 0 \\ 0 \\ -1 \\ 1 \\ 1 \end{array} \right) \ .$$

State 3 specific solution vectors.

To illustrate the back-substitution method in this example, note that the matrix form shown just before the final row-reduced echelon form satisfies the requirements that the pivots are established and the elements below them are zero. The equations corresponding to this matrix are

$$\begin{aligned} x_1 + x_2 - x_3 + 2x_4 - x_5 + 5x_6 &= 6 \\ x_4 + x_6 &= 2 \\ x_5 - x_6 &= 1 \quad . \end{aligned}$$

Solve the third equation for x_5 to obtain $x_5 = 1 + x_6$.

Solve the second equation for x_4 to obtain $x_4 = 2 - x_6$.

Substitute these values of x_4 and x_5 into the first equation and solve for x_1 to obtain

$$x_1 = 3 - x_2 + x_3 - 2x_6 \quad .$$

These equations are the same as those obtained from the row-reduced echelon form.

3.5 Solving Systems of Linear Equations: A Summary

- Write the augmented matrix for the system.

- Row-reduce it enough to establish the pivot element and to make all the elements in the column below each pivot zero.

- If a row occurs with all elements zero except the last element, then the system is inconsistent and the solution set is empty.

- Otherwise,

  Use the pivots in reverse order (bottom to top) to eliminate non-zero column elements above them to obtain the row-reduced echelon form.

  If the number of pivots is one less than the number of columns of the augmented matrix, then the solution set contains exactly one member (the *single solution case*).

  Otherwise, the pivot variables are expressed in terms of one or more free variables which are set equal to arbitrary parameters (the *underdetermined case*) and the solution set contains an infinite number of members corresponding to the infinite choice of values for these parameters.

3.6 Exercises

1. Show that each of the following systems is inconsistent

a) $0a = 2$

b) $$\begin{aligned} 2a + b &= 0 \\ 2a + b &= 1 \end{aligned}$$

c) $$\begin{aligned} a - b &= 0 \\ b - a &= 1 \end{aligned}$$

d) $$\begin{aligned} a + b - c &= 1 \\ -a - b + c &= 2 \end{aligned}$$

e) $$\begin{aligned} a - b &= 1 \\ a + b &= 0 \\ -a + b &= 0 \end{aligned}$$

f) $$\begin{aligned} 2a - 3b + c &= 1 \\ a + b - c &= 0 \\ a - 4b + 2c &= -1 \end{aligned}$$

g) $$\begin{aligned} x_1 + x_2 + x_3 + x_4 &= 1 \\ x_1 - x_2 + x_3 - x_4 &= 0 \\ 2x_1 &= 1 \\ 2x_2 - 2x_3 + 2x_4 &= 1 \\ 3x_1 + x_2 - x_3 + x_4 &= 1 \ . \end{aligned}$$

2. Confirm that each of the following systems has a coefficient matrix with rank equal to the number of unknowns. Find the unique solution for each system.

a) $-3w = -6$

b) $$\begin{aligned} 2w + x &= 1 \\ w + x &= 0 \end{aligned}$$

c) $$\begin{aligned} 3w - x &= 0 \\ 2w + 5x &= 0 \end{aligned}$$

d) $$\begin{aligned} 2w - 4x &= 18 \\ -w + 3x &= -11 \end{aligned}$$

e) $$\begin{aligned} 5w - 3x + 6y &= 0 \\ w + x - 2y &= 0 \\ -2w - x + y &= 0 \end{aligned}$$

f) $$\begin{aligned} 2w - x + 2y &= -1 \\ w - 2x + 3y &= -4 \\ 3w + 2x + 2y &= 3 \end{aligned}$$

g) $$\begin{aligned} -2w + x - 3y &= 1 \\ 2w - 2x + y &= -3 \\ w + x - y &= -3 \end{aligned}$$

h) $$\begin{aligned} x_1 - 2x_2 + x_3 + 2x_4 &= 0 \\ x_1 + x_2 - 3x_3 + x_4 &= -3 \\ -2x_1 + x_2 - x_3 - 2x_4 &= -1 \\ 3x_1 + 2x_2 - x_3 + 2x_4 &= 0 \end{aligned}$$

i) $$\begin{aligned} x_1 + x_2 + x_3 + x_4 &= 0 \\ x_1 - x_3 &= 0 \\ 2x_1 + x_2 &= 0 \\ x_2 - x_4 &= 0 \end{aligned}$$

j) $$\begin{aligned} -2x_1 + 2x_2 - x_3 &= -4 \\ 5x_2 + x_3 - 2x_4 &= -3 \\ 2x_1 - x_2 + 2x_3 + x_4 &= 5 \\ 2x_2 + x_3 - x_4 &= 0 \ . \end{aligned}$$

3. Each of the following systems of equations has a coefficient matrix with rank less than the number of unknowns. Find all solutions.

a) $3y + 3z = 3$ **b)** $5y - 10z = 0$ **c)** $2x + 2y - 6z = 4$

d)
$$\begin{aligned} x - 3y + z &= 4 \\ -2x - 19y + 3z &= -3 \end{aligned}$$

e)
$$\begin{aligned} x + y + z &= 0 \\ -4x + 2y - z &= -3 \\ -5x + y - 2z &= -3 \end{aligned}$$

f)
$$\begin{aligned} x_1 + x_2 - x_3 + x_4 &= -1 \\ 2x_1 + 3x_2 - 3x_3 + 2x_4 &= -4 \\ -x_1 - 2x_2 + 2x_3 - x_4 &= 3 \end{aligned}$$

g)
$$\begin{aligned} x_1 + x_2 + x_3 + x_4 &= 0 \\ 2x_1 - 2x_2 - x_3 + x_4 - x_5 &= 0 \\ -x_1 + 5x_2 + x_3 - 2x_4 - x_5 &= 0 \\ -x_1 - 7x_2 - 2x_3 + x_4 + 2x_5 &= 0 \\ x_1 + 7x_2 + 2x_3 - x_4 - 2x_5 &= 0 \\ -2x_1 - 8x_2 - 2x_3 + x_4 + 3x_5 &= 0 \ . \end{aligned}$$

4. Find all solutions of the following systems of equations.

a)
$$\begin{aligned} 3x + y - 2z &= 0 \\ -2x - y + z &= 1 \\ -x - 2y - z &= -4 \end{aligned}$$

b)
$$\begin{aligned} 2x - y + z &= 5 \\ x + 2y - z &= 0 \\ -x + y + 2z &= -3 \\ x - y - z &= 3 \end{aligned}$$

c)
$$\begin{aligned} y + 3z &= 2 \\ 2y + 6z &= 4 \\ -3y - 9z &= -6 \end{aligned}$$

d)
$$\begin{aligned} 2x - y + 3z &= -1 \\ x + y + z &= 0 \\ -x + 2y - 3z &= 2 \end{aligned}$$

e)
$$\begin{aligned} x_1 + x_2 + 3x_3 + x_5 &= 0 \\ x_1 - x_2 - x_3 - 2x_4 - x_5 &= 0 \\ -x_1 + x_2 + x_3 + 2x_4 + x_5 &= 0 \\ x_1 + 2x_2 + x_3 + x_4 - 2x_5 &= 0 \end{aligned}$$

f)
$$\begin{aligned} x + y + z &= 2 \\ x - y - z &= 2 \\ x + y - z &= 0 \ . \end{aligned}$$

5. The general equation for a plane in $\mathcal{R}^3$ is $ax+by+cz+d = 0$. Determine an equation for the plane containing the three points $(0,0,-1)$, $(1,1,4)$, and $(-1,1,0)$.

6. Find the general solution of each of the following equations or else show that it is inconsistent.

a) $\begin{pmatrix} 3 & 6 \\ 6 & 12 \end{pmatrix} \begin{pmatrix} x_1 \\ x_2 \end{pmatrix} = \begin{pmatrix} 3 \\ 12 \end{pmatrix}$ **b)** $\begin{pmatrix} 3 & 6 \\ 6 & 12 \end{pmatrix} \begin{pmatrix} x_1 \\ x_2 \end{pmatrix} = \begin{pmatrix} -3 \\ -6 \end{pmatrix}$

c) $$\begin{pmatrix} 1 & 2 & -4 \\ -1 & 1 & 1 \\ 1 & 5 & -7 \end{pmatrix} \begin{pmatrix} x_1 \\ x_2 \\ x_3 \end{pmatrix} = \begin{pmatrix} 1 \\ -1 \\ 2 \end{pmatrix}$$

d) $$\begin{pmatrix} 1 & 2 & -4 \\ -1 & 1 & 1 \\ 1 & 5 & -7 \end{pmatrix} \begin{pmatrix} x_1 \\ x_2 \\ x_3 \\ x_4 \end{pmatrix} = \begin{pmatrix} 3 \\ -3 \\ 3 \end{pmatrix}$$

e) $$\begin{pmatrix} 1 & 1 & 0 & 1 \\ 1 & 1 & 1 & 0 \\ 1 & 2 & 1 & 3 \\ 2 & 1 & 0 & -1 \end{pmatrix} \begin{pmatrix} x_1 \\ x_2 \\ x_3 \\ x_4 \end{pmatrix} = \begin{pmatrix} 5 \\ 2 \\ 10 \\ 2 \end{pmatrix}$$

f) $$\begin{pmatrix} 1 & 1 & 0 & 1 \\ 1 & 1 & 1 & 0 \\ 1 & 2 & 1 & 3 \\ 2 & 1 & 0 & -1 \end{pmatrix} \begin{pmatrix} x_1 \\ x_2 \\ x_3 \\ x_4 \end{pmatrix} = \begin{pmatrix} 1 \\ 2 \\ 2 \\ 1 \end{pmatrix}.$$

7. The ***augmented matrix*** for a system of equations in variables

$$(x_1, x_2, x_3, x_4, x_5, x_6, x_7)$$

has the following row reduced echelon form

$$R = \left(\begin{array}{ccccccc|c} 0 & 1 & 0 & 0 & 3 & 0 & 7 & 2 \\ 0 & 0 & 0 & 1 & 5 & 0 & 6 & 8 \\ 0 & 0 & 0 & 0 & 0 & 1 & 4 & 9 \end{array}\right).$$

Write out the ***general solution*** of the system.

4 The Determinant of a Matrix

Outline of Key Ideas

Basic concepts:

the determinant is a number

the determinant of a 2×2 matrix

Cramer's Rule for a 2×2 system

the determinant of a 3×3 matrix.

Two methods for calculating a determinant:

the method of Laplace Expansion

the method of row reduction to triangular form.

Additional properties and applications of determinants.

The Determinant of a Matrix

4.1 Basic Concepts

The determinant of a square $n \times n$ matrix A is a number p.

We write $\det A = p$. For example, if A is the 2×2 matrix

$$A = \begin{pmatrix} 1 & 2 \\ 3 & 4 \end{pmatrix}$$

then $\det A = -2$, as we shall soon discover.

To motivate the rule for obtaining the number which is the determinant of a 2×2 matrix, consider the system of two linear equations in the two variables x and y:

$$a_{11}x + a_{12}y = b_1 \tag{4.1}$$

$$a_{21}x + a_{22}y = b_2 \quad . \tag{4.2}$$

Consider the case that all of the constants a_{11}, a_{12}, a_{21}, a_{22} are non-zero. Multiply Equation (4.1) by a_{22} and Equation (4.2) by $-a_{12}$ and add the resulting equations. This produces an equation in which there is no y term:

$$(a_{11}a_{22} - a_{12}a_{21})x = a_{22}b_1 - a_{12}b_2 \quad . \tag{4.3}$$

Now multiply Equation (4.1) by $-a_{21}$ and Equation (4.2) by a_{11} and add the resulting equations. This produces an equation in which there is no x term:

$$(a_{11}a_{22} - a_{12}a_{21})y = a_{11}b_2 - a_{21}b_1 \quad . \tag{4.4}$$

Notice that the quantity $a_{11}a_{22} - a_{12}a_{21}$ occurs as the coefficient of the variable in each of the equations (4.3) and (4.4). If this quantity is non-zero, division by it immediately produces the ***unique solution*** to the system

$$x = \frac{a_{22}b_1 - a_{12}b_2}{a_{11}a_{22} - a_{12}a_{21}} \tag{4.5}$$

$$y = \frac{-a_{21}b_1 + a_{11}b_2}{a_{11}a_{22} - a_{12}a_{21}} \quad . \tag{4.6}$$

If the quantity $a_{11}a_{22} - a_{12}a_{21}$ is zero, then there ***does not exist a unique solution*** to the system. The system either has no solution or has infinitely many solutions. Further information is needed to determine which case applies.

Notice that the coefficient matrix for the original system of equations is

$$A = \begin{pmatrix} a_{11} & a_{12} \\ a_{21} & a_{22} \end{pmatrix} \tag{4.7}$$

and that its elements are involved in the quantity $a_{11}a_{22} - a_{12}a_{21}$, the value of which determines whether or not the system has a unique solution.

Definition 4.1 The ***determinant*** of the 2×2 matrix (4.7) is given by $a_{11}a_{22}-a_{12}a_{21}$ and is called a determinant of order 2.

We can write this in any one of the following ways:

$$\det A = \det \begin{pmatrix} a_{11} & a_{12} \\ a_{21} & a_{22} \end{pmatrix} = \begin{vmatrix} a_{11} & a_{12} \\ a_{21} & a_{22} \end{vmatrix} = a_{11}a_{22} - a_{12}a_{21} \quad . \tag{4.8}$$

Example 4.1 Compute $\begin{vmatrix} 3 & -4 \\ 6 & -2 \end{vmatrix}$.

Solution.

$$\begin{vmatrix} 3 & -4 \\ 6 & -2 \end{vmatrix} = (3)(-2) - (-4)(6) = 18 \quad .$$

Be careful not to confuse $\begin{pmatrix} a_{11} & a_{12} \\ a_{21} & a_{22} \end{pmatrix}$ *(a matrix)* with $\begin{vmatrix} a_{11} & a_{12} \\ a_{21} & a_{22} \end{vmatrix}$ *(a number)*. Determinants are defined ***only*** for square matrices.

We now reconsider the analysis of the system of equations (4.1) and (4.2) with this new terminology. The important number in that analysis is the determinant of the coefficient matrix A. The case $\det A$ not equal 0 corresponds to the system having a unique solution, (*the single solution case in Chapter 3*), while $\det A = 0$ corresponds to the solution set being either empty (*the inconsistent system*) or containing infinitely many members (*the consistent system with the rank less than the number of unknowns*).

When $\det A \neq 0$, the determinant may be used to rewrite equations (4.5) and (4.6) as:

$$x = \frac{\begin{vmatrix} b_1 & a_{12} \\ b_2 & a_{22} \end{vmatrix}}{\begin{vmatrix} a_{11} & a_{12} \\ a_{21} & a_{22} \end{vmatrix}} \tag{4.9}$$

$$y = \frac{\begin{vmatrix} a_{11} & b_1 \\ a_{21} & b_2 \end{vmatrix}}{\begin{vmatrix} a_{11} & a_{12} \\ a_{21} & a_{22} \end{vmatrix}} \quad . \tag{4.10}$$

This is called ***Cramer's Rule*** for solving a consistent system of two linear equations in two unknowns. Cramer's Rule for larger consistent systems will be discussed in Chapter 5.

We have given the rule needed to compute the determinant of a 2×2 matrix in (4.8). The rule for computing the determinant of a 3×3 matrix is given by

$$\det A = \begin{vmatrix} a_{11} & a_{12} & a_{13} \\ a_{21} & a_{22} & a_{23} \\ a_{31} & a_{32} & a_{33} \end{vmatrix} = \tag{4.11}$$

$$a_{11}a_{22}a_{33} + a_{12}a_{23}a_{31} + a_{13}a_{21}a_{32} - a_{31}a_{22}a_{13} - a_{32}a_{23}a_{11} - a_{33}a_{21}a_{12}$$

and is called a determinant of order 3.

4.2 The Method of Laplace Expansion

The rules for computing determinants of larger square matrices could be written in a similar way, but they quickly become more cumbersome as the size of the matrix increases. A more convenient approach involves a method called ***Laplace Expansion along a row or column***. We will use the 3×3 case to illustrate this.

First we need some terminology. Select any (i,j) position of the matrix A and form a corresponding submatrix S_{ij} of A by eliminating all the elements from the row and column passing through the (i,j) position (they contain a_{ij} in their intersection). For example, in (4.11) look at the $(1,2)$ position then S_{12} is given by

$$S_{12} = \begin{pmatrix} a_{21} & a_{23} \\ a_{31} & a_{33} \end{pmatrix} \quad . \tag{4.12}$$

The (i,j) ***minor*** of A is written as M_{ij} and is defined by

$$M_{ij} = \det S_{ij} \tag{4.13}$$

and the (i,j) ***cofactor*** of A, written as A_{ij}, is defined by

$$A_{ij} = (-1)^{i+j} M_{ij} \quad . \tag{4.14}$$

In the case of (4.11)

$$A_{12} = (-1)^{1+2} \det S_{12} = -\begin{vmatrix} a_{21} & a_{23} \\ a_{31} & a_{33} \end{vmatrix} = -(a_{21}a_{33} - a_{31}a_{23}) = a_{31}a_{23} - a_{21}a_{33} \quad .$$

Note that there is a cofactor associated with each position of a matrix.

Now the Laplace Expansion Method may be described. Select any row or column along which to expand. For each position in the selected row or column, form the product of the element with its own cofactor. The determinant of the matrix is the sum of these products. For example, choose the second column in (4.11), then

$$(4.15) \qquad \begin{vmatrix} a_{11} & a_{12} & a_{13} \\ a_{21} & a_{22} & a_{23} \\ a_{31} & a_{32} & a_{33} \end{vmatrix}$$

$$\begin{aligned} &= a_{12}\left(-\begin{vmatrix} a_{21} & a_{23} \\ a_{31} & a_{33} \end{vmatrix}\right) + a_{22}\begin{vmatrix} a_{11} & a_{13} \\ a_{31} & a_{33} \end{vmatrix} + a_{32}\left(-\begin{vmatrix} a_{11} & a_{13} \\ a_{21} & a_{23} \end{vmatrix}\right) \\ &= a_{12}A_{12} + a_{22}A_{22} + a_{32}A_{32} \quad . \end{aligned}$$

Calculate the determinants of the 2×2 matrices in (4.15) and show that the result agrees with (4.11).

Example 4.2 Compute the determinant of

$$\begin{pmatrix} -1 & 2 & 1 \\ 1 & 0 & 3 \\ 4 & 6 & 7 \end{pmatrix} .$$

1. By expansion along the first row

$$\begin{vmatrix} -1 & 2 & 1 \\ 1 & 0 & 3 \\ 4 & 6 & 7 \end{vmatrix} = (-1)\begin{vmatrix} 0 & 3 \\ 6 & 7 \end{vmatrix} + (2)\left(-\begin{vmatrix} 1 & 3 \\ 4 & 7 \end{vmatrix}\right) + (1)\begin{vmatrix} 1 & 0 \\ 4 & 6 \end{vmatrix} =$$

$$a_{11}A_{11} + a_{12}A_{12} + a_{13}A_{13} = (-1)(-18) + (2)(5) + (1)(6) = 34 .$$

2. By expansion along the second column:

$$\begin{vmatrix} -1 & 2 & 1 \\ 1 & 0 & 3 \\ 4 & 6 & 7 \end{vmatrix} = (2)\left(-\begin{vmatrix} 1 & 3 \\ 4 & 7 \end{vmatrix}\right) + (0)\begin{vmatrix} -1 & 1 \\ 4 & 7 \end{vmatrix} + (6)\left(-\begin{vmatrix} -1 & 1 \\ 1 & 3 \end{vmatrix}\right) =$$

$$a_{12}A_{12} + a_{22}A_{22} + a_{32}A_{32} = (2)(5) + (0)(-11) + (6)(4) = 34 .$$

Expanding along the kth row of the $n \times n$ matrix A, the Laplace expansion method gives

$$\det A = a_{k1}A_{k1} + \cdots + a_{kn}A_{kn} = \sum_{i=1}^{n} a_{ki}A_{ki} \tag{4.16}$$

for the nth order determinant $\det A$ of the $n \times n$ matrix A. Notice that each of the cofactors A_{ki} involves an $(n-1)^{\text{th}}$ order determinant of an $(n-1) \times (n-1)$ matrix. Thus the method gives a perfectly adequate theoretical procedure in principle, but in practice the calculation is generally prohibitively long and subject to serious round off error. The occurrence of many zero elements in the matrix can make the calculations considerably shorter. Consider the following example.

Example 4.3 Find the determinant of

$$A = \begin{pmatrix} 3 & 0 & 4 & 2 \\ 1 & -1 & 0 & 1 \\ 0 & -2 & 0 & -1 \\ 1 & 0 & 1 & 0 \end{pmatrix} .$$

Solution.

Choose a row or column containing the maximum number of zero elements. For example, choose the fourth row. Then omitting from the sum the terms with zero we obtain

$$\det A = 1\left(-\begin{vmatrix} 0 & 4 & 2 \\ -1 & 0 & 1 \\ -2 & 0 & -1 \end{vmatrix}\right) + 1\left(-\begin{vmatrix} 3 & 0 & 2 \\ 1 & -1 & 1 \\ 0 & -2 & -1 \end{vmatrix}\right) .$$

Now expand about the second column of the first cofactor to get

$$-\begin{vmatrix} 0 & 4 & 2 \\ -1 & 0 & 1 \\ -2 & 0 & -1 \end{vmatrix} = (-1)(4)\left(-\begin{vmatrix} -1 & 1 \\ -2 & -1 \end{vmatrix}\right) = (-1)4(-3) = 12$$

and for the second 3×3, use the first column to get

$$-\begin{vmatrix} 3 & 0 & 2 \\ 1 & -1 & 1 \\ 0 & -2 & -1 \end{vmatrix} = (-1)\left(3\begin{vmatrix} -1 & 1 \\ -2 & -1 \end{vmatrix} + 1\left(-\begin{vmatrix} 0 & 2 \\ -2 & -1 \end{vmatrix}\right)\right) =$$

$$-(3(3) + 1(-4)) = -5 \ .$$

Then $\det A = (1)(12) + 1(-5) = 7$.

4.3 The Method of Row Reduction to Triangular Form

A ***triangular matrix*** is a square matrix with all of its elements below (or *above*) the diagonal having value zero. Its determinant is especially easy to calculate.

Theorem 4.1. *The determinant of a triangular matrix is the product of the diagonal elements.*

$$\det A = a_{11}a_{22}\ldots a_{nn} = \prod_{i=1}^{n} a_{ii} \quad . \tag{4.17}$$

Example 4.4 Direct calculation in the 2×2 case demonstrates the theorem.

$$\begin{vmatrix} a_{11} & a_{12} \\ 0 & a_{22} \end{vmatrix} = a_{11}a_{22} \quad .$$

Example 4.5 For a 3×3 matrix, expand about the first column to get:

$$\begin{vmatrix} a_{11} & a_{12} & a_{13} \\ 0 & a_{22} & a_{23} \\ 0 & 0 & a_{33} \end{vmatrix} = a_{11} \begin{vmatrix} a_{22} & a_{23} \\ 0 & a_{33} \end{vmatrix}$$

and do it once more to get

$$a_{11} \begin{vmatrix} a_{22} & a_{23} \\ 0 & a_{33} \end{vmatrix} = a_{11}a_{22}a_{33} \quad .$$

Example 4.6 Using the theorem,

$$\begin{vmatrix} 1 & 0 & 0 & 0 & 0 & 0 \\ 52 & -1 & 0 & 0 & 0 & 0 \\ 31 & 0 & 3 & 0 & 0 & 0 \\ -16 & 41.5 & -61 & -\pi & 0 & 0 \\ 102 & \pi & 22 & 14 & -2 & 0 \\ -5 & -e & 4 & -1 & 7 & 0.5 \end{vmatrix} = 1(-1)3(-\pi)(-2)(0.5) = -3\pi \quad .$$

Since we already know how to reduce any square matrix to triangular form using the three elementary row operations, we could calculate the determinant of the original matrix by calculating the determinant of its row-equivalent triangular form if we knew how the two determinants are related. The following theorems are concerned with the effect of each of the three elementary row operations on the determinant of the matrix.

Theorem 4.2. *Suppose two rows of matrix A are interchanged to produce matrix B. Then*

$$\det B = -\det A \quad .$$

Example 4.7 The determinant of the matrix in Example 4.2 was found to be 34. Construct matrix B from that matrix by interchanging the first two rows. A calculation of its determinant will produce the value -34.

Example 4.8 The 2×2 case can be easily demonstrated.

$$\begin{vmatrix} a_{11} & a_{12} \\ a_{21} & a_{22} \end{vmatrix} = a_{11}a_{22} - a_{12}a_{21}$$

and

$$\begin{vmatrix} a_{21} & a_{22} \\ a_{11} & a_{12} \end{vmatrix} = a_{21}a_{12} - a_{11}a_{22} = -(a_{11}a_{22} - a_{12}a_{21}) = -\begin{vmatrix} a_{11} & a_{12} \\ a_{21} & a_{22} \end{vmatrix} \quad .$$

Theorem 4.3. *Multiply one row of matrix A by a scalar r to produce matrix B. Then*

$$\det B = r \det A \quad .$$

Example 4.9 The determinant of the matrix in Example 4.2 is 34. Multiply its first row by 2 to produce the matrix B. A calculation of its determinant will produce a value of 68.

Example 4.10 In the 2×2 case

$$\begin{vmatrix} ra_{11} & ra_{12} \\ a_{21} & a_{22} \end{vmatrix} = ra_{11}a_{22} - ra_{12}a_{21} = r(a_{11}a_{22} - a_{12}a_{21}) = r\begin{vmatrix} a_{11} & a_{12} \\ a_{21} & a_{22} \end{vmatrix} \quad .$$

Theorem 4.4. *Produce matrix B from the matrix A by replacing any one row of A by that row added to a multiple of any other row of A. Then*

$$\det B = \det A \quad .$$

Example 4.11 From Example 4.2, $\begin{vmatrix} -1 & 2 & 1 \\ 1 & 0 & 3 \\ 4 & 6 & 7 \end{vmatrix} = 34$.

Add 4 times row 1 to row 3 and compute the determinant by row 2 expansion to obtain

$$\begin{vmatrix} -1 & 2 & 1 \\ 1 & 0 & 3 \\ 0 & 14 & 11 \end{vmatrix} = 1\left(-\begin{vmatrix} 2 & 1 \\ 14 & 11 \end{vmatrix}\right) + 3\left(-\begin{vmatrix} -1 & 2 \\ 0 & 14 \end{vmatrix}\right) = 1(-8) + 3(14) = 34 \quad .$$

Example 4.12 In the 2×2 case,

$$\begin{vmatrix} a_{11} + ra_{21} & a_{12} + ra_{22} \\ a_{21} & a_{22} \end{vmatrix} = (a_{11} + ra_{21})a_{22} - (a_{12} + ra_{22})a_{21} =$$

$$a_{11}a_{22} + ra_{21}a_{22} - a_{12}a_{21} - ra_{21}a_{22} = a_{11}a_{22} - a_{12}a_{21} = \begin{vmatrix} a_{11} & a_{12} \\ a_{21} & a_{22} \end{vmatrix} \quad .$$

As a consequence of these three theorems, if matrices A and B are row equivalent, then their determinants are very simply related. One could in fact employ a sequence of only elementary row operations of type 3 to reduce the original matrix A to a triangular form B. Then $\det A = \det B$ by Theorem 4.4. However, use of elementary row operations of types 1 and 2 can sometimes save steps or simplify arithmetic in the reduction process. If they are employed, their simple effects on the determinant according to Theorem 4.2 and 4.3 have to be taken into account.

Example 4.13 Using only type 3 elementary row operations,

$$\begin{vmatrix} 1 & 2 & 3 \\ -1 & 2 & 1 \\ 4 & 1 & -1 \end{vmatrix} = \begin{vmatrix} 1 & 2 & 3 \\ 0 & 4 & 4 \\ 0 & -7 & -13 \end{vmatrix} = \begin{vmatrix} 1 & 2 & 3 \\ 0 & 4 & 4 \\ 0 & 0 & -6 \end{vmatrix} = 1(4)(-6) = -24 \quad .$$

Note that if these operations are combined with expansions about simplified first columns we have

$$\begin{vmatrix} 1 & 2 & 3 \\ -1 & 2 & 1 \\ 4 & 1 & -1 \end{vmatrix} = \begin{vmatrix} 1 & 2 & 3 \\ 0 & 4 & 4 \\ 0 & -7 & -13 \end{vmatrix} = \begin{vmatrix} 4 & 4 \\ -7 & -13 \end{vmatrix} = \begin{vmatrix} 4 & 4 \\ 0 & -6 \end{vmatrix} = -24 \quad .$$

Example 4.14 Using only type 1 elementary row operations,

$$\begin{vmatrix} 0 & 0 & 2 & -3 \\ 0 & 0 & 0 & 5 \\ -2 & 6 & 1 & 9 \\ 0 & 3 & -8 & 1 \end{vmatrix} = -\begin{vmatrix} -2 & 6 & 1 & 9 \\ 0 & 0 & 0 & 5 \\ 0 & 0 & 2 & -3 \\ 0 & 3 & -8 & 1 \end{vmatrix} =$$

$$\begin{vmatrix} -2 & 6 & 1 & 9 \\ 0 & 3 & -8 & 1 \\ 0 & 0 & 2 & -3 \\ 0 & 0 & 0 & 5 \end{vmatrix} = (-2)(3)(2)(5) = -60 \quad .$$

Example 4.15 Using all types,

$$\begin{vmatrix} 3 & 0 & 4 & 2 \\ 1 & -1 & 0 & 1 \\ 0 & -2 & 0 & -1 \\ 1 & 0 & 1 & 0 \end{vmatrix} = -\begin{vmatrix} 1 & 0 & 1 & 0 \\ 1 & -1 & 0 & 1 \\ 0 & -2 & 0 & -1 \\ 3 & 0 & 4 & 2 \end{vmatrix} = -\begin{vmatrix} 1 & 0 & 1 & 0 \\ 0 & -1 & -1 & 1 \\ 0 & -2 & 0 & -1 \\ 0 & 0 & 1 & 2 \end{vmatrix} =$$

$$\begin{vmatrix} 1 & 0 & 1 & 0 \\ 0 & 1 & 1 & -1 \\ 0 & -2 & 0 & -1 \\ 0 & 0 & 1 & 2 \end{vmatrix} = \begin{vmatrix} 1 & 0 & 1 & 0 \\ 0 & 1 & 1 & -1 \\ 0 & 0 & 2 & -3 \\ 0 & 0 & 1 & 2 \end{vmatrix} = -\begin{vmatrix} 1 & 0 & 1 & 0 \\ 0 & 1 & 1 & -1 \\ 0 & 0 & 1 & 2 \\ 0 & 0 & 2 & -3 \end{vmatrix} =$$

$$-\begin{vmatrix} 1 & 0 & 1 & 0 \\ 0 & 1 & 1 & -1 \\ 0 & 0 & 1 & 2 \\ 0 & 0 & 0 & -7 \end{vmatrix} = (-1)(-7) = 7 \quad .$$

Again note that if these operations are combined with expansions about first columns we have

$$\begin{vmatrix} 3 & 0 & 4 & 2 \\ 1 & -1 & 0 & 1 \\ 0 & -2 & 0 & -1 \\ 1 & 0 & 1 & 0 \end{vmatrix} = -\begin{vmatrix} 1 & 0 & 1 & 0 \\ 0 & -1 & -1 & 1 \\ 0 & -2 & 0 & -1 \\ 0 & 0 & 1 & 2 \end{vmatrix} = -\begin{vmatrix} -1 & -1 & 1 \\ -2 & 0 & -1 \\ 0 & 1 & 2 \end{vmatrix}$$

$$= -\begin{vmatrix} -1 & -1 & 1 \\ 0 & 2 & -3 \\ 0 & 1 & 2 \end{vmatrix} = (-1)(-1)\begin{vmatrix} 2 & -3 \\ 1 & 2 \end{vmatrix} = -\begin{vmatrix} 1 & 2 \\ 2 & -3 \end{vmatrix} = -\begin{vmatrix} 1 & 2 \\ 0 & -7 \end{vmatrix} = 7 \quad .$$

We have seen three approaches to the computation of determinants, the Laplace Expansion Method, the reduction to triangular form technique, and an equivalent technique using Gaussian elimination on the first column and a cofactor expansion with respect to the first column.

We will call this last technique ***pivotal condensation***. The principal idea of the technique is to zero out every entry in the first column below a leading non-zero pivot and then reduce the nth order determinant to the product of the pivot and its cofactor, an $(n-1)^{\text{th}}$ order determinant. The effect of the first column on the determinant is to reduce or condense it to the product of the pivot times its cofactor. It is not essential that the first column be used in pivotal condensation as will be illustrated in the next section.

4.4 Some Additional Properties of Determinants and an Efficient Method to Compute Them

The following list summarizes other useful properties of determinants.
Let A and B be $n \times n$ matrices, and let A^T be the transpose of A. Then

(4.18)

If A has a zero row (column), then $\det A = 0$.

(4.19)

If any row (column) of A is a multiple of another row (column), then $\det A = 0$.

(4.20)
$$\det AB = (\det A)(\det B) \quad .$$

(4.21)
$$\det A^T = \det A \quad .$$

We could start with a matrix A and take its transpose A^T and then use a sequence of elementary row operations to compute $\det A^T$. But because of (4.21) these elementary row operations on $\det A^T$ are equivalent to elementary column operations on $\det A$, since the rows of A^T are the columns of A. Thus any convenient combination of row and column operations can be used to compute $\det A$. A very efficient numerical method to compute $\det A$ is to use ***pivotal condensation*** with respect to a convenient non-zero entry in A. Either the row or column containing the element is used. The order or size of the determinants that need to be computed are systematically reduced down to order 2.

The determinants from Examples 4.13, 4.14, and 4.15 will each be computed in two ways.

Example 4.16

a) For $A = \begin{pmatrix} 1 & 2 & 3 \\ -1 & 2 & 1 \\ 4 & 1 & -1 \end{pmatrix}$ the element $a_{23} = 1$ will be used as a 1st pivot. Zeroing out the other entries of the 3rd column we obtain

$$\det A = \begin{vmatrix} 4 & -4 & 0 \\ -1 & 2 & 1 \\ 3 & 3 & 0 \end{vmatrix} = -\begin{vmatrix} 4 & -4 \\ 3 & 3 \end{vmatrix} = -(4)(3)\begin{vmatrix} 1 & -1 \\ 1 & 1 \end{vmatrix} = (-12)(1+1) = -24$$

If we zero out the other entries of the 2nd row of A we obtain

$$\det A = \begin{vmatrix} 4 & -4 & 3 \\ 0 & 0 & 1 \\ 3 & 3 & -1 \end{vmatrix} = -\begin{vmatrix} 4 & -4 \\ 3 & 3 \end{vmatrix} = -(12+12) = -24 \quad .$$

b) For $A = \begin{pmatrix} 0 & 0 & 2 & -3 \\ 0 & 0 & 0 & 5 \\ -2 & 6 & 1 & 9 \\ 0 & 3 & -8 & 1 \end{pmatrix}$ its 2nd row is already in the desired form so

$$\det A = 5\begin{vmatrix} 0 & 0 & 2 \\ -2 & 6 & 1 \\ 0 & 3 & -8 \end{vmatrix} = (5)(2)\begin{vmatrix} -2 & 6 \\ 0 & 3 \end{vmatrix} = (10)(-6) = -60 \quad .$$

To check this result use the element $a_{33} = 1$ and zero out the other entries in the 3rd column to obtain

$$\det A = \begin{vmatrix} 4 & -12 & 0 & -21 \\ 0 & 0 & 0 & 5 \\ -2 & 6 & 1 & 9 \\ -16. & 51 & 0 & 73 \end{vmatrix} = \begin{vmatrix} 4 & -12 & -21 \\ 0 & 0 & 5 \\ -16 & 51 & 73 \end{vmatrix} = -5\begin{vmatrix} 4 & -12 \\ -16 & 51 \end{vmatrix}$$

$$= -20\begin{vmatrix} 1 & -3 \\ -16 & 51 \end{vmatrix} = -20\begin{vmatrix} 1 & -3 \\ 0 & 3 \end{vmatrix} = -60 \ .$$

c) For $A = \begin{pmatrix} 3 & 0 & 4 & 2 \\ 1 & -1 & 0 & 1 \\ 0 & -2 & 0 & -1 \\ 1 & 0 & 1 & 0 \end{pmatrix}$ use the element $a_{24} = 1$ as a 1st pivot. Zeroing out the other entries of the 4th column we obtain

$$\det A = \begin{vmatrix} 1 & 2 & 4 & 0 \\ 1 & -1 & 0 & 1 \\ 1 & -3 & 0 & 0 \\ 1 & 0 & 1 & 0 \end{vmatrix} = \begin{vmatrix} 1 & 2 & 4 \\ 1 & -3 & 0 \\ 1 & 0 & 1 \end{vmatrix}$$

$$= \begin{vmatrix} -3 & 2 & 0 \\ 1 & -3 & 0 \\ 1 & 0 & 1 \end{vmatrix} = \begin{vmatrix} -3 & 2 \\ 1 & -3 \end{vmatrix} = 9 - 2 = 7 \ .$$

If we zero out the other entries of the 2nd row of A we obtain

$$\det A = \begin{vmatrix} 1 & 2 & 4 & 2 \\ 0 & 0 & 0 & 1 \\ 1 & -3 & 0 & -1 \\ 1 & 0 & 1 & 0 \end{vmatrix} = \begin{vmatrix} 1 & 2 & 4 \\ 1 & -3 & 0 \\ 1 & 0 & 1 \end{vmatrix} =$$

$$\begin{vmatrix} -3 & 2 & 4 \\ 1 & -3 & 0 \\ 0 & 0 & 1 \end{vmatrix} = \begin{vmatrix} -3 & 2 \\ 1 & -3 \end{vmatrix} = 9 - 2 = 7 \ .$$

In the beginning of this section, we saw that the role of the determinant of the coefficient matrix of a system of two linear equations was important in deciding the nature of the solution set. We also saw that Cramer's Rule was expressed entirely in terms of determinants. This will be extended to larger systems in the next section. Additional uses of the determinant are presented in the following examples.

4.5 Some Applications of Determinants

Example 4.17 The definition of the cross product of two vectors $\mathbf{a} = \langle a_1, a_2, a_3 \rangle$ and $\mathbf{b} = \langle b_1, b_2, b_3 \rangle$ in $\mathcal{R}^3$ is conveniently represented as

$$\mathbf{a} \times \mathbf{b} = \begin{vmatrix} \mathbf{i} & \mathbf{j} & \mathbf{k} \\ a_1 & a_2 & a_3 \\ b_1 & b_2 & b_3 \end{vmatrix} = (a_2 b_3 - a_3 b_2)\mathbf{i} - (a_1 b_3 - a_3 b_1)\mathbf{j} + (a_1 b_2 - a_2 b_1)\mathbf{k} \quad .$$

Example 4.18 The volume V of a parallelepiped in $\mathcal{R}^3$ with edge vectors $\mathbf{a} = \langle a_1, a_2, a_3 \rangle$, $\mathbf{b} = \langle b_1, b_2, b_3 \rangle$, and $\mathbf{c} = \langle c_1, c_2, c_3 \rangle$ is given by $V = |d|$ where

$$\mathbf{d} = \mathbf{a} \bullet \mathbf{b} \times \mathbf{c} = \begin{vmatrix} a_1 & a_2 & a_3 \\ b_1 & b_2 & b_3 \\ c_1 & c_2 & c_3 \end{vmatrix} \quad .$$

Example 4.19 The area A of a triangle in the plane with vertices (x_1, y_1), (x_2, y_2), and (x_3, y_3) is given by $A = \frac{1}{2}|d|$ where

$$d = \begin{vmatrix} x_1 & y_1 & 1 \\ x_2 & y_2 & 1 \\ x_3 & y_3 & 1 \end{vmatrix} \quad .$$

For another important application of the determinant, see Theorem 5.1.

4.6 Exercises

1. Evaluate the following determinants:

a) $\begin{vmatrix} 3 & -1 \\ 2 & 4 \end{vmatrix}$ **b)** $\begin{vmatrix} 1 & 2 \\ 3 & 5 \end{vmatrix}$ **c)** $\begin{vmatrix} \ln 2 & \ln 3 \\ 3 & 2 \end{vmatrix}$

d) $\begin{vmatrix} \cos(t) & -\sin(t) \\ \sin(t) & \cos(t) \end{vmatrix}$ **e)** $\begin{vmatrix} \sec(x) & -1 \\ -1 & \sec(x) \end{vmatrix}$ **f)** $\begin{vmatrix} -5 & -1 \\ 2 & -3 \end{vmatrix}$.

2. Evaluate the following determinants:

a) $\begin{vmatrix} 3 & 0 & 0 \\ 4 & 5 & 0 \\ -1 & 2 & 6 \end{vmatrix}$ **b)** $\begin{vmatrix} 2 & 2 & 2 \\ 1 & 1 & 1 \\ -1 & -1 & -1 \end{vmatrix}$ **c)** $\begin{vmatrix} 1 & 2 & -1 \\ 3 & 1 & 2 \\ 1 & 0 & 2 \end{vmatrix}$

d) $\begin{vmatrix} 2 & -1 & 3 \\ 1 & 1 & 6 \\ 0 & 2 & -2 \end{vmatrix}$ **e)** $\begin{vmatrix} 1 & 0 & 1 & 1 \\ 0 & 0 & 1 & -1 \\ 0 & 1 & 0 & -1 \\ -1 & -1 & 0 & 0 \end{vmatrix}$ **f)** $\begin{vmatrix} -1 & 2 & 0 & 0 \\ 1 & -1 & 1 & -1 \\ 1 & 2 & 0 & 1 \\ 0 & 3 & 1 & 2 \end{vmatrix}$.

3. For what values of t does the matrix have a determinant equal to zero?

a) $\begin{pmatrix} t & -2 \\ 3 & 12 \end{pmatrix}$ **b)** $\begin{pmatrix} 1-t & 0 \\ 3 & -3-t \end{pmatrix}$ **c)** $\begin{pmatrix} 1-t & 1 & 1 \\ 1 & 2-t & 0 \\ 1 & 0 & 2-t \end{pmatrix}$

d) $\begin{pmatrix} t & 0 & 2 \\ 0 & t-1 & 0 \\ 2 & 0 & t \end{pmatrix}$.

4. **a)** Show that $\begin{vmatrix} x_1 & y_1 & 1 \\ x_2 & y_2 & 1 \\ x & y & 1 \end{vmatrix} = 0$ is an equation of the form $ax + by + c = 0$ and therefore represents a line.

b) Show that (x_1, y_1) and (x_2, y_2) lie on this line.

c) Use **a)** to find the equation of the line containing $(-1, 1)$ and $(2, -1)$.

5. Using Example 4.19, find the area of the triangle with vertices at $(1, -1)$, $(3, 6)$, and $(2, 2)$.

6. Using Example 4.18, find the volume of the parallelepiped with edge vectors $\langle 1, 0, 2\rangle$, $\langle 0, 1, 1\rangle$, and $\langle -1, 1, 0\rangle$.

7. **a)** Show that if A is a 2×2 matrix, then $\det(-A) = \det A$.

b) Show that if A is a 3×3 matrix, then $\det(-A) = -\det A$.

8. Using Cramer's Rule, solve the given system of equations.

a) $\begin{aligned} x + 2y &= 5 \\ 3x - y &= 1 \end{aligned}$ **b)** $\begin{aligned} 2x + 3y &= 2 \\ 4x - 9y &= -1 \end{aligned}$ **c)** $\begin{aligned} x - 3y &= 0 \\ 2x - 9y &= 3 \end{aligned}$

d) $$\begin{aligned} x + y &= -1 \\ x - y &= 1 \, . \end{aligned}$$

9. Find $\begin{vmatrix} -1 & 0 & 0 & 1 \\ 0 & 1 & -1 & 0 \\ 1 & -1 & -2 & 0 \\ 0 & 0 & 1 & 1 \end{vmatrix}$

a) by using the Laplace expansion method.

b) by using row reduction to triangular form.

c) by using pivotal condensation starting with the 1st pivot $a_{14} = 1$ and zeroing out the rest of the 4th column.

d) Repeat **c)** except zero out the rest of the 1st row.

10. Let $A = \begin{pmatrix} -1 & 3 & 1 & 3 \\ 1 & 1 & -2 & 1 \\ 2 & 1 & 1 & 1 \\ -1 & -2 & 1 & 3 \end{pmatrix}$. Find the cofactors A_{23} and A_{31}.

11. Expand each of the following determinants in three ways: first with respect to the cofactors of the ***second column***, then with respect to the cofactors of the ***last row***, and then using elementary row and column operations in a pivotal condensation.

a) $\begin{vmatrix} 2 & 5 \\ 3 & 1 \end{vmatrix}$ **b)** $\begin{vmatrix} 1 & 2 & 1 \\ 1 & 0 & 2 \\ 0 & -2 & 3 \end{vmatrix}$ **c)** $\begin{vmatrix} 2 & 1 & -1 \\ -1 & 1 & 2 \\ 1 & 2 & 1 \end{vmatrix}$

d) $\begin{vmatrix} 1 & 1 & 1 \\ 1 & -1 & 2 \\ 1 & 1 & 4 \end{vmatrix}$ **e)** $\begin{vmatrix} 1 & 1 & 1 \\ 2 & -2 & 1 \\ 8 & -8 & 1 \end{vmatrix}$ **f)** $\begin{vmatrix} 0 & 2 & 1 & 1 \\ 1 & 0 & -1 & 0 \\ 0 & 1 & 0 & -1 \\ 2 & 3 & 0 & 1 \end{vmatrix}$.

5 The Inverse of a Matrix

Outline of Key Ideas

Basic concepts:

definition of the inverse of a matrix

the determinant test for the existence of an inverse.

Two methods for calculating the inverse matrix:

the row reduction method

the adjoint matrix method.

Solving systems of linear equations with the inverse matrix:

Cramer's rule.

Additional properties of the matrix inverse operation.

Inverse of a Matrix

5.1 Basic Concepts

If a is a non-zero real number, there is another non-zero real number b such that

$$ab = 1 \quad . \tag{5.1}$$

The numbers a and b are said to be inverses of each other with respect to multiplication. For example, 3 and $\frac{1}{3}$ are inverses of each other (*with respect to multiplication*), as is $\pi \approx 3.14159$ and $\frac{1}{\pi} \approx 0.318310$. The test of whether or not two numbers are inverses is Equation (5.1), i.e. whether or not their product is 1. Note that zero is the only real number without an inverse.

Does matrix multiplication permit inverses? A similar approach does define multiplicative inverses for matrices, but the existence of an inverse for matrices is rarer than for real numbers. In particular, matrix conformability is needed for the multiplication to be defined and even then the product is not in general commutative. The resulting complications are avoided by ***defining an inverse which only*** $n \times n$ ***matrices can possess***. Non-square matrices can only possess one-sided inverses which are beyond the scope of this discussion.

Definition 5.1 Two $n \times n$ matrices A and B are said to be ***inverses*** of each other if and only if

$$AB = BA = I_n \quad . \tag{5.2}$$

Example 5.1 If the two real numbers a and b are inverses of each other, then the 1×1 matrices $A = [a]$ and $B = [b]$ are inverses of each other.

$$\begin{aligned} AB &= [a][b] = [ab] = [1] = I_1 \\ BA &= [b][a] = [ba] = [1] = I_1 \quad . \end{aligned}$$

Example 5.2 Let $A = \begin{pmatrix} 2 & 3 \\ 1 & 2 \end{pmatrix}$ and $B = \begin{pmatrix} 2 & -3 \\ -1 & 2 \end{pmatrix}$ then explicit computation shows that $BA = AB = \begin{pmatrix} 1 & 0 \\ 0 & 1 \end{pmatrix}$ so that A and B are inverses of each other.

A common notation is to write A^{-1} instead of B. (Note that A^{-1} does ***not*** mean $1/A$ or A to the negative one power.) We call A^{-1} the ***inverse of*** A or the matrix inverse of A. In this notation

$$AA^{-1} = A^{-1}A = I_n \quad .$$

It is a fact that for square matrices $AA^{-1} = I_n$ if and only if $AA^{-1} = I_n$. As a consequence, ***we need only demonstrate one or the other equality, but not both.***

Example 5.3 Show that

$$A = \begin{pmatrix} 1 & 0 & 1 \\ 2 & 1 & 1 \\ 3 & 2 & 2 \end{pmatrix}$$

and

$$B = \begin{pmatrix} 0 & 2 & -1 \\ -1 & -1 & 1 \\ 1 & -2 & 1 \end{pmatrix}$$

are inverses of each other.

Solution.

$$\text{Compute } AB = \begin{pmatrix} 1 & 0 & 1 \\ 2 & 1 & 1 \\ 3 & 2 & 2 \end{pmatrix} \begin{pmatrix} 0 & 2 & -1 \\ -1 & -1 & 1 \\ 1 & -2 & 1 \end{pmatrix} = \begin{pmatrix} 1 & 0 & 0 \\ 0 & 1 & 0 \\ 0 & 0 & 1 \end{pmatrix} = I_3 \quad .$$

Not every square matrix has an inverse matrix. For example, a matrix with at least one zero row or column has no inverse matrix.

Example 5.4 Suppose the 2×2 matrix A has a zero first row. Try to construct a 2×2 matrix B to be the inverse of A. Then B must do the following job ($\star$ indicates an arbitrary real number):

$$AB = \begin{pmatrix} 0 & 0 \\ \star & \star \end{pmatrix} \begin{pmatrix} c & \star \\ d & \star \end{pmatrix} = I_2 = \begin{pmatrix} 1 & 0 \\ 0 & 1 \end{pmatrix} \quad .$$

In particular, the $(1,1)$ element I_2 must be produced by multiplication of the 1st row of A times the 1st column of B, i.e.

$$0 = 0c + 0d = 1 \quad .$$

But there are no real numbers c and d which will make this equation true. Hence A has no inverse.

How can we know whether a specific $n \times n$ matrix A does or does not have an inverse? The following theorem contains a very simple and useful application of determinants to answer this question.

Theorem 5.1. *The $n \times n$ matrix A has an inverse matrix A^{-1} if and only if* $\det A \neq 0$.

An invertible matrix (one which has an inverse) is called a ***nonsingular*** matrix, and a matrix which has no inverse is called a ***singular*** matrix. Theorem 5.1 states that a singular matrix has a zero determinant and a nonsingular matrix has a non-zero determinant.

Example 5.5 For

$$A = \begin{pmatrix} 2 & 3 \\ 1 & 2 \end{pmatrix} ,$$

$\det A = 1$. Thus A has an inverse matrix. In fact its inverse was stated in Example 5.2.

Example 5.6 For

$$A = \begin{pmatrix} 1 & 0 & 1 \\ 2 & 1 & 1 \\ 3 & 2 & 2 \end{pmatrix} ,$$

$\det A = 1$. Thus A is nonsingular. Its inverse matrix was stated in Example 5.3.

Example 5.7 For

$$A = \begin{pmatrix} 1 & 1 \\ 2 & 2 \end{pmatrix} ,$$

$\det A = 0$. Thus A is a singular matrix. There is no inverse for A.

5.2 Two Methods for Calculating the Inverse Matrix

1. **Row Reduction Method**

   Due to Theorem 5.1, we know that if $\det A \neq 0$, then A^{-1} exists. But how do we actually find it? The next theorem gives the most efficient numerical technique.

   Theorem 5.2. *If* $\det A \neq 0$, *the row-reduced echelon form of the matrix* $[A|I_n]$ *is the matrix* $[I_n|A^{-1}]$. *Furthermore, the row-reduced echelon form of A is I_n so A has rank n when it is invertible.*

Example 5.8 Find A^{-1} for $A = \begin{pmatrix} 2 & 3 \\ 1 & 2 \end{pmatrix}$.

Solution.

Since $\det A = 1$, we need to find the row reduced echelon form of

$$[A|I_2] = \left(\begin{array}{cc|cc} 2 & 3 & 1 & 0 \\ 1 & 2 & 0 & 1 \end{array}\right) \sim \left(\begin{array}{cc|cc} 1 & 2 & 0 & 1 \\ 2 & 3 & 1 & 0 \end{array}\right) \sim \left(\begin{array}{cc|cc} 1 & 2 & 0 & 1 \\ 0 & -1 & 1 & -2 \end{array}\right) \sim$$

$$\left(\begin{array}{cc|cc} 1 & 2 & 0 & 1 \\ 0 & 1 & -1 & 2 \end{array}\right) \sim \left(\begin{array}{cc|cc} 1 & 0 & 2 & -3 \\ 0 & 1 & -1 & 2 \end{array}\right) = [I_2|A^{-1}]$$

Thus $A^{-1} = \begin{pmatrix} 2 & -3 \\ -1 & 2 \end{pmatrix}$. Example 5.2 showed this to be the correct inverse.

Example 5.9 Find A^{-1} for $A = \begin{pmatrix} 1 & 0 & 1 \\ 2 & 1 & 1 \\ 3 & 2 & 2 \end{pmatrix}$. Since $\det A = 1$, we row reduce $[A|I_3]$.

$$[A|I_3] = \left(\begin{array}{ccc|ccc} 1 & 0 & 1 & 1 & 0 & 0 \\ 2 & 1 & 1 & 0 & 1 & 0 \\ 3 & 2 & 2 & 0 & 0 & 1 \end{array}\right) \sim \left(\begin{array}{ccc|ccc} 1 & 0 & 1 & 1 & 0 & 0 \\ 0 & 1 & -1 & -2 & 1 & 0 \\ 0 & 2 & -1 & -3 & 0 & 1 \end{array}\right) \sim$$

$$\left(\begin{array}{ccc|ccc} 1 & 0 & 1 & 1 & 0 & 0 \\ 0 & 1 & -1 & -2 & 1 & 0 \\ 0 & 0 & 1 & 1 & -2 & 1 \end{array}\right) \sim \left(\begin{array}{ccc|ccc} 1 & 0 & 0 & 0 & 2 & -1 \\ 0 & 1 & 0 & -1 & -1 & 1 \\ 0 & 0 & 1 & 1 & -2 & 1 \end{array}\right) = [I_3|A^{-1}] \ .$$

Hence $A^{-1} = \begin{pmatrix} 0 & 2 & -1 \\ -1 & -1 & 1 \\ 1 & -2 & 1 \end{pmatrix}$. This is the correct inverse (*Example* 5.3).

2. **Adjoint Matrix Method**

Another method for finding A^{-1} involves the ***adjoint*** of A. If A is an $n \times n$ matrix, the adjoint of A is constructed by

1) finding all n^2 cofactors of A,

2) putting them into a matrix of cofactors,

and

3) Transposing the resulting matrix of cofactors. The resulting matrix is the adjoint of A and is denoted by $\operatorname{adj} A$.

Using the symbol A_{ij} for the (i,j) cofactor of A (see Equation (4.14)), we have

$$\operatorname{adj} A = \begin{pmatrix} A_{11} & \cdots & A_{1n} \\ A_{21} & \cdots & A_{2n} \\ \vdots & & \vdots \\ A_{n1} & \cdots & A_{nn} \end{pmatrix}^T = \begin{pmatrix} A_{11} & \cdots & A_{n1} \\ A_{12} & \cdots & A_{n2} \\ \vdots & & \vdots \\ A_{1n} & \cdots & A_{nn} \end{pmatrix} .$$

Then A^{-1} is related to $\operatorname{adj} A$ by the following theorem.

Theorem 5.3. *If* $\det A \neq 0$ *then*

$$A^{-1} = \frac{1}{\det A} \operatorname{adj} A \quad .$$

Example 5.10 Use the $\operatorname{adj} A$ to find A^{-1} for $A = \begin{pmatrix} 2 & 3 \\ 1 & 2 \end{pmatrix}$.

Solution.

$$\operatorname{adj} A = \begin{pmatrix} 2 & -1 \\ -3 & 2 \end{pmatrix}^T = \begin{pmatrix} 2 & -3 \\ -1 & 2 \end{pmatrix}$$

Since $\det A = 1$,

$$A^{-1} = \begin{pmatrix} 2 & -3 \\ -1 & 2 \end{pmatrix}$$

which agrees with Example 5.8.

Example 5.11 Use the adjoint method to find A^{-1} for

$$A = \begin{pmatrix} 1 & 0 & 1 \\ 2 & 1 & 1 \\ 3 & 2 & 2 \end{pmatrix} .$$

Solution.

Since $\det A = 1$, we have

$$A^{-1} = \operatorname{adj} A = \begin{pmatrix} 0 & -1 & 1 \\ 2 & -1 & -2 \\ -1 & 1 & 1 \end{pmatrix}^T = \begin{pmatrix} 0 & 2 & -1 \\ -1 & -1 & 1 \\ 1 & -2 & 1 \end{pmatrix}$$

which agrees with Example 5.9.

For matrices of small size, the row reduction method and the adjoint method use a comparable number of arithmetic operations. However, for matrices of large size, the row reduction method uses considerably fewer arithmetic operations than the adjoint method. Hence the row reduction method is generally preferred for ***computation*** of A^{-1}. On the other hand, the adjoint method is valuable for theoretical purposes.

5.3 The Inverse Matrix Applied to the Solution of Systems of Linear Equations (Cramer's Rule)

The inverse matrix is useful for solving a consistent system of equations for which the coefficient matrix A has rank n. To see how, first consider solving the algebraic equation

$$ax = b \quad . \tag{5.3}$$

We naturally solve for x by dividing both sides of the equation by a, i.e. multiplying both sides by a^{-1} (assuming $a \neq 0$)

$$x = a^{-1}b = \frac{b}{a} \quad . \tag{5.4}$$

In a similar way, we can solve the matrix equation representing a system of equations

$$A\mathbf{x} = \mathbf{b} \tag{5.5}$$

for the solution vector $\mathbf{x}$ if we have A^{-1} available. Multiplying both sides from the left by A^{-1}, we have

$$A^{-1}A\mathbf{x} = I\mathbf{x} = \mathbf{x} = A^{-1}\mathbf{b} \quad . \tag{5.6}$$

Example 5.12 Find the solution vectors to each of the following systems:

1)

$$\begin{array}{rcl} u_1 + \quad\quad + u_3 & = & 1 \\ 2u_1 + u_2 + u_3 & = & 1 \\ 3u_1 + 2u_2 + 2u_3 & = & 1 \end{array} ,$$

2)

$$\begin{array}{rcl} v_1 \quad\quad\quad + v_3 & = & 1 \\ 2v_1 + v_2 + v_3 & = & 0 \\ 3v_1 + 2v_2 + 2v_3 & = & -1 \end{array} ,$$

3)

$$\begin{array}{rcl} w_1 \quad\quad\quad + w_3 & = & 0 \\ 2w_1 + w_2 + w_3 & = & 0 \\ 3w_1 + 2w_2 + 2w_3 & = & 0 \end{array} .$$

Solution.

In terms of matrices these can be written as

$$A\mathbf{u} = \mathbf{b}, \quad A\mathbf{v} = \mathbf{c}, \quad \text{and} \quad A\mathbf{w} = \mathbf{0} \quad \text{where}$$

$$A = \begin{pmatrix} 1 & 0 & 1 \\ 2 & 1 & 1 \\ 3 & 2 & 2 \end{pmatrix}, \quad \mathbf{u} = \begin{pmatrix} u_1 \\ u_2 \\ u_3 \end{pmatrix}, \quad \mathbf{v} = \begin{pmatrix} v_1 \\ v_2 \\ v_3 \end{pmatrix}, \quad \mathbf{w} = \begin{pmatrix} w_1 \\ w_2 \\ w_3 \end{pmatrix},$$

$$\mathbf{b} = \begin{pmatrix} 1 \\ 1 \\ 1 \end{pmatrix}, \quad \& \quad \mathbf{c} = \begin{pmatrix} 1 \\ 0 \\ -1 \end{pmatrix}.$$

We have previously computed A^{-1} for this A in Examples 5.9 and 5.11. The solutions to the systems are consequently given by

$$\mathbf{u} = A^{-1}\mathbf{b} = \begin{pmatrix} 1 \\ -1 \\ 0 \end{pmatrix}, \quad \mathbf{v} = A^{-1}\mathbf{c} = \begin{pmatrix} 1 \\ -2 \\ 0 \end{pmatrix}, \ \& \ \mathbf{w} = A^{-1}\mathbf{0} = \mathbf{0} = \begin{pmatrix} 0 \\ 0 \\ 0 \end{pmatrix} .$$

Note that this method works ***only*** for the case where the system is consistent and the coefficient matrix A has rank n, because only then does A^{-1} exist. We know that there is exactly one solution for such a system and this method gives it directly. A more explicit way of writing this solution directly in terms of determinants is obtained by using the adjoint form of A^{-1}. The details are not treated here, but the result, known as ***Cramer's Rule***, is the following

Theorem 5.4 (Cramer's Rule). *Let* $A\mathbf{x} = \mathbf{b}$ *be a linear system with an* $n \times n$ *invertible coefficient matrix* A *and* $\mathbf{x} = (x_1, \ldots, x_n)^T$. *Let* N_i *be the* $n \times n$ *matrix obtained by replacing the* i^{th} *column of* A *with the column vector* $\mathbf{b}$ *and leaving all the other columns of* A *unchanged,* $i = 1, \ldots, n$. *The components of* $\mathbf{x}$ *are given by*

$$x_i = \frac{\det N_i}{\det A}, \quad i = 1, 2, \ldots, n \quad .$$

The method of Gaussian Elimination is much more efficient for computing the solution of a large system of equations, but the more explicit form offered by Cramer's Rule can be convenient for a small system of rank n such as in the following example.

Example 5.13 Use Cramer's Rule to solve the first system in Example 5.12.

Solution.

$$x_1 = \frac{\begin{vmatrix} 1 & 0 & 1 \\ 1 & 1 & 1 \\ 1 & 2 & 2 \end{vmatrix}}{\begin{vmatrix} 1 & 0 & 1 \\ 2 & 1 & 1 \\ 3 & 2 & 2 \end{vmatrix}} = \frac{1}{1} = 1 \quad x_2 = \frac{\begin{vmatrix} 1 & 1 & 1 \\ 2 & 1 & 1 \\ 3 & 1 & 2 \end{vmatrix}}{\begin{vmatrix} 1 & 0 & 1 \\ 2 & 1 & 1 \\ 3 & 2 & 2 \end{vmatrix}} = \frac{-1}{1} = -1 \quad x_3 = \frac{\begin{vmatrix} 1 & 0 & 1 \\ 2 & 1 & 1 \\ 3 & 2 & 1 \end{vmatrix}}{\begin{vmatrix} 1 & 0 & 1 \\ 2 & 1 & 1 \\ 3 & 2 & 2 \end{vmatrix}} = \frac{0}{1} = 0.$$

Thus the solution vector is $\begin{pmatrix} 1 \\ -1 \\ 0 \end{pmatrix}$ as obtained in Example 5.12.

5.4 Some Additional Properties of the Matrix Inverse Operation

Some useful properties of matrix inverses are (A and B are nonsingular $n \times n$ matrices, c is a non-zero real number):

$$
\begin{aligned}
&(5.7) \qquad & (A^{-1})^{-1} &= A \\
&(5.8) \qquad & I_n^{-1} &= I_n \\
&(5.9) \qquad & (A^T)^{-1} &= (A^{-1})^T \\
&(5.10) \qquad & (AB)^{-1} &= B^{-1}A^{-1} \\
&(5.11) \qquad & (cA)^{-1} &= c^{-1}A^{-1} \quad .
\end{aligned}
$$

5.5 Exercises

1. For which of the following matrices does an inverse matrix exist?

a) $\begin{pmatrix} 0 \end{pmatrix}$ **b)** $\begin{pmatrix} 2 \end{pmatrix}$ **c)** $\begin{pmatrix} 1 & 2 \end{pmatrix}$ **d)** $\begin{pmatrix} 1 & 2 \\ 2 & 1 \end{pmatrix}$

e) $\begin{pmatrix} 1 & 2 \\ -1 & -2 \end{pmatrix}$ **f)** $\begin{pmatrix} 0 & 1 \\ 1 & 0 \end{pmatrix}$ **g)** $\begin{pmatrix} 1 & 0 \\ 0 & 1 \end{pmatrix}$ **h)** $\begin{pmatrix} 1 & -1 \\ 2 & 2 \\ 3 & 1 \end{pmatrix}$

i) $\begin{pmatrix} 1 & 1 & 2 \\ 3 & -1 & 0 \\ -1 & 3 & 4 \end{pmatrix}$ **j)** $\begin{pmatrix} 1 & 0 & 1 \\ -1 & 1 & 2 \\ 0 & 1 & -1 \end{pmatrix}$ **k)** $\begin{pmatrix} 1 & 0 & 1 & 1 \\ -1 & 0 & 0 & 1 \\ 1 & 1 & 1 & 0 \end{pmatrix}$

l) $\begin{pmatrix} 0 & 0 & 0 & 1 \\ 0 & 0 & 1 & 0 \\ 0 & 1 & 0 & 0 \\ 1 & 0 & 0 & 0 \end{pmatrix}$ **m)** $\begin{pmatrix} 1 & 0 & 0 & 0 \\ 0 & 1 & 0 & 0 \\ 0 & 0 & 1 & 0 \\ 0 & 0 & 0 & 1 \end{pmatrix}$.

2. Where possible, find the inverse of each matrix in Problem 1.

3. Are the matrices A and B inverses of each other?

$$A = \begin{pmatrix} 1 & 0 & -1 & 1 \\ 1 & 0 & 0 & 1 \\ 1 & 1 & -1 & 0 \\ 0 & -1 & 1 & 0 \end{pmatrix} \quad \& \quad B = \begin{pmatrix} 0 & 0 & 1 & 1 \\ -1 & 1 & 0 & -1 \\ -1 & 1 & 0 & 0 \\ 0 & 1 & -1 & -1 \end{pmatrix} .$$

4. State the condition for which

$$A = \begin{pmatrix} a & b \\ c & d \end{pmatrix}$$

has an inverse and find A^{-1}.

5. State the condition for which

$$A = \begin{pmatrix} a & 0 & 0 \\ 0 & b & 0 \\ 0 & 0 & c \end{pmatrix}$$

has an inverse and find A^{-1}.

6. Solve all the following systems of equations using the fact that if

$$A = \begin{pmatrix} 1 & 0 & 1 \\ 2 & 1 & 1 \\ 3 & 2 & 2 \end{pmatrix} \quad \text{then} \quad A^{-1} = \begin{pmatrix} 0 & 2 & -1 \\ -1 & -1 & 1 \\ 1 & -2 & 1 \end{pmatrix} .$$

a)
$$\begin{aligned} x + z &= 1 \\ 2x + y + z &= 0 \\ 3x + 2y + 2z &= 0 \end{aligned}$$

b)
$$\begin{aligned} x + z &= -1 \\ 2x + y + z &= 2 \\ 3x + 2y + 2z &= -1 \end{aligned}$$

c)
$$\begin{aligned} x + z &= 1 \\ 2x + y + z &= 1 \\ 3x + 2y + 2z &= 2 \end{aligned}$$

d)
$$\begin{aligned} x + z &= 0 \\ 2x + y + z &= 0 \\ 3x + 2y + 2z &= 0 \end{aligned}$$

e)
$$\begin{aligned} x + z &= 2 \\ 2x + y + z &= 5 \\ 3x + 2y + 2z &= -3 \end{aligned} .$$

7. Solve the system

$$\begin{aligned} x + z &= 2 \\ 2x + y + z &= 5 \\ 3x + 2y + 2z &= -3 \end{aligned}$$

using Cramer's Rule.

8. Use the adjoint method to find A^{-1} if

$$A = \begin{pmatrix} 1 & 0 & 1 \\ -1 & 1 & 0 \\ 0 & 1 & 0 \end{pmatrix} .$$

9. Find A^{-1} if $A = A(\beta) = \begin{pmatrix} \cos\beta & -\sin\beta \\ \sin\beta & \cos\beta \end{pmatrix}$. Do you see any simple relationship between A^{-1} and $A = A(\beta)$ in this case?

10. **a)** Try to use Cramer's rule to solve

$$\begin{aligned} x + y &= 3 \\ 2x + 2y &= 6 \end{aligned} .$$

Explain why it doesn't work.

b) Use another method to solve the system.

11. If each one of the three elementary row operations is applied to an identity matrix the resulting matrix is called an elementary matrix. If A is an $m \times n$ matrix and P is an $m \times m$ elementary matrix then PA is the matrix obtained from A by the application of the elementary row operation corresponding to P. Find the following elementary matrices P and verify PA corresponds to the indicated elementary row operation.

a) P adds β times the 1st row of a 3×3 matrix to its 3rd row,

b) P interchanges the 2nd and 4th rows of a 4×4 matrix,

c) P multiplies the 2nd row of a 4×4 matrix by $\alpha \neq 0$.

12. Find the inverses of the elementary matrices in Exercise 11.

13. Find A and A^{-1} if $A = P_1 P_2$ where P_1 and P_2 are the following elementary matrices

a) $P_1 = \begin{pmatrix} 1 & 2 \\ 0 & 1 \end{pmatrix}, \quad P_2 = \begin{pmatrix} 1 & 0 \\ 3 & 1 \end{pmatrix}$

b) $P_1 = \begin{pmatrix} 1 & 3 & 0 \\ 0 & 1 & 0 \\ 0 & 0 & 1 \end{pmatrix}, \quad P_2 = \begin{pmatrix} 1 & 0 & 0 \\ 0 & 1 & 0 \\ 0 & 2 & 1 \end{pmatrix}$

c) $P_1 = \begin{pmatrix} 1 & 0 & 4 \\ 0 & 1 & 0 \\ 0 & 0 & 1 \end{pmatrix}, \quad P_2 = \begin{pmatrix} 0 & 1 & 0 \\ 1 & 0 & 0 \\ 0 & 0 & 1 \end{pmatrix}$

d) $P_1 = \begin{pmatrix} 1 & 0 & 0 & 0 \\ 0 & 1 & 3 & 0 \\ 0 & 0 & 1 & 0 \\ 0 & 0 & 0 & 1 \end{pmatrix}, \quad P_2 = \begin{pmatrix} 1 & 0 & 0 & 0 \\ 0 & 1 & 0 & 0 \\ 5 & 0 & 1 & 0 \\ 0 & 0 & 0 & 1 \end{pmatrix}.$

14. Compute the adjoint adj A of each of the following matrices $A = [a_{ij}]$ and verify that $A(\operatorname{adj} A) = (\operatorname{adj} A)A = (\det A)I$.

a) $\begin{pmatrix} 2 & 0 \\ 0 & 7 \end{pmatrix}$ **b)** $\begin{pmatrix} 0 & 2 \\ 3 & 0 \end{pmatrix}$ **c)** $\begin{pmatrix} 0 & 3 \\ -5 & 7 \end{pmatrix}$. **d)** $\begin{pmatrix} 2 & 3 \\ -5 & 7 \end{pmatrix}$

e) $\begin{pmatrix} 1 & 1 & 2 \\ 2 & -1 & 0 \\ 3 & 2 & -1 \end{pmatrix}$ **f)** $\begin{pmatrix} 1 & 2 & 0 \\ -1 & 0 & 1 \\ 0 & 2 & 0 \end{pmatrix}$ **g)** $\begin{pmatrix} 2 & 0 & 0 \\ 0 & 3 & 5 \\ 0 & 2 & 4 \end{pmatrix}$.

15. Verify that the following are correct

a) $\begin{pmatrix} 2 & 3 \\ 4 & 5 \end{pmatrix}^{-1} = \frac{1}{2}\begin{pmatrix} -5 & 3 \\ 4 & -2 \end{pmatrix}$ **b)** $\begin{pmatrix} 0 & 0 & 4 \\ 0 & 3 & 0 \\ 2 & 0 & 0 \end{pmatrix}^{-1} = \begin{pmatrix} 0 & 0 & \frac{1}{2} \\ 0 & \frac{1}{3} & 0 \\ \frac{1}{4} & 0 & 0 \end{pmatrix}$

c) $\begin{pmatrix} 1 & 2 & 2 \\ 2 & 1 & -2 \\ 2 & -2 & 1 \end{pmatrix}^{-1} = \frac{1}{9}\begin{pmatrix} 1 & 2 & 2 \\ 2 & 1 & -2 \\ 2 & -2 & 1 \end{pmatrix}$

d) $\begin{pmatrix} 1 & 0 & 0 & 0 \\ 0 & 0 & 0 & 5 \\ 0 & 1 & 0 & 3 \\ 0 & 0 & 1 & 0 \end{pmatrix}^{-1} = \begin{pmatrix} 1 & 0 & 0 & 0 \\ 0 & \frac{-3}{5} & 1 & 0 \\ 0 & 0 & 0 & 1 \\ 0 & \frac{1}{5} & 0 & 0 \end{pmatrix}$

e) $\begin{pmatrix} 1 & \alpha & 0 \\ 0 & 1 & \beta \\ 0 & 0 & 1 \end{pmatrix}^{-1} = \begin{pmatrix} 1 & -\alpha & \alpha\beta \\ 0 & 1 & -\beta \\ 0 & 0 & 1 \end{pmatrix}$.

16. Solve the following equations

a) $\begin{pmatrix} 2 & 3 \\ 4 & 5 \end{pmatrix}\begin{pmatrix} x_1 \\ x_2 \end{pmatrix} = \begin{pmatrix} 4 \\ 6 \end{pmatrix}$

b) $\begin{pmatrix} 0 & 0 & 4 \\ 0 & 3 & 0 \\ 2 & 0 & 0 \end{pmatrix}\begin{pmatrix} x_{11} & x_{12} \\ x_{21} & x_{22} \\ x_{31} & x_{32} \end{pmatrix} = \begin{pmatrix} -16 & 20 \\ 3 & -6 \\ 2 & 4 \end{pmatrix}$

c) $\begin{pmatrix} 1 & 2 & 2 \\ 2 & 1 & -2 \\ 2 & -2 & 1 \end{pmatrix}\begin{pmatrix} x_{11} & x_{12} & x_{13} & x_{14} \\ x_{21} & x_{22} & x_{23} & x_{24} \\ x_{31} & x_{32} & x_{33} & x_{34} \end{pmatrix} = \begin{pmatrix} 9 & -9 & 27 & 9 \\ -9 & 27 & -18 & 9 \\ 18 & 9 & -9 & 9 \end{pmatrix}$

d) $\begin{pmatrix} 1 & 0 & 0 & 0 \\ 0 & 0 & 0 & 5 \\ 0 & 1 & 0 & 3 \\ 0 & 0 & 1 & 0 \end{pmatrix}\begin{pmatrix} x_1 \\ x_2 \\ x_3 \\ x_4 \end{pmatrix} = \begin{pmatrix} 1 \\ 5 \\ -4 \\ 1 \end{pmatrix}$.

17. If $A = [a_{ij}]^{n \times n}$ is invertible, show that $(A^2)^{-1} = (A^{-1})^2$ and $(A^3)^{-1} = (A^{-1})^3$.

18. If $A = [a_{ij}]^{n \times n}$ and $B = [b_{ij}]^{n \times n}$ are invertible, show that $[(AB)^{-1}]^T = (A^{-1})^T(B^{-1})^T$, and verify the result for

a) $A = \begin{pmatrix} 5 & 4 \\ 4 & 3 \end{pmatrix} \quad B = \begin{pmatrix} 6 & 5 \\ 5 & 4 \end{pmatrix}$

b) $A = \begin{pmatrix} 1 & -1 & 0 \\ -1 & 2 & -1 \\ 0 & -1 & 2 \end{pmatrix} \quad B = \begin{pmatrix} 1 & 2 & 1 \\ 2 & 1 & 0 \\ 1 & 0 & 1 \end{pmatrix}$

c) $A = \begin{pmatrix} 1 & 0 & 2 \\ 0 & 1 & -1 \\ 0 & 0 & 1 \end{pmatrix} \quad B = \begin{pmatrix} 1 & 0 & 0 \\ 3 & 1 & 0 \\ 1 & 0 & 1 \end{pmatrix}$.

6 Orthogonal Matrices and Changes of Coordinates

Outline of Key Ideas

The definition of an orthogonal matrix: $Q^{-1} = Q^T$

equivalent characterizations:

columns (*rows*) have unit length

and

columns (*rows*) are mutually orthogonal.

Coordinates and changes or coordinates:

basis vectors and linear independence

general changes of coordinates: $\mathbf{v} = A\mathbf{v}'$, $\mathbf{v}' = A^{-1}\mathbf{v}$.

Changes of coordinates with orthogonal matrices:

$\mathbf{v} = Q\mathbf{v}'$, $\mathbf{v}' = Q^T\mathbf{v}$

the columns of Q are the coordinate columns of the new primed coordinate axes

orthonormal sets of vectors are linearly independent

the general form of an arbitrary 2×2 orthogonal matrix

rotation matrices: Q orthogonal and $\det Q = 1$

right handed and left handed sets of orthonormal basis vectors

two and three dimensional examples.

Orthogonal Matrices and Changes of Coordinates

The matrix Q given by

$$Q = \frac{1}{\sqrt{2}} \begin{pmatrix} 1 & -1 \\ 1 & 1 \end{pmatrix} \tag{6.1}$$

is one example of an orthogonal matrix. In this chapter, we will study such matrices and their use in changes or coordinates.

We first note an important property of the matrix Q in (6.1).

$$QQ^T = Q^T Q = I \tag{6.2}$$

or

$$\frac{1}{\sqrt{2}} \begin{pmatrix} 1 & -1 \\ 1 & 1 \end{pmatrix} \frac{1}{\sqrt{2}} \begin{pmatrix} 1 & 1 \\ -1 & 1 \end{pmatrix} = \frac{1}{\sqrt{2}} \begin{pmatrix} 1 & 1 \\ -1 & 1 \end{pmatrix} \frac{1}{\sqrt{2}} \begin{pmatrix} 1 & -1 \\ 1 & 1 \end{pmatrix} = \begin{pmatrix} 1 & 0 \\ 0 & 1 \end{pmatrix} .$$

By the definition of Q^{-1} this shows that Q^T (*the transpose of* Q) is the inverse of Q or

$$Q^{-1} = Q^T \quad . \tag{6.3}$$

This property is characteristic of all orthogonal matrices and is used to define them.

Definition 6.1 An $n \times n$ invertible matrix Q is called an ***orthogonal matrix*** if and only if $Q^{-1} = Q^T$.

Note that inverses of orthogonal matrices are very easy to find, since we only need to take transposes. Finding inverses by the standard methods of Chapter 5 is generally quite complicated.

6.1 An Equivalent Characterization of an Orthogonal Matrix

For convenience express the matrix Q in (6.1) in terms of its columns. Let $\mathbf{q}_1$ be the first column of Q and $\mathbf{q}_2$ be the second column. Then $Q = [\mathbf{q}_1, \mathbf{q}_2]$ where

$$\mathbf{q}_1 = \frac{1}{\sqrt{2}} \begin{pmatrix} 1 \\ 1 \end{pmatrix} \qquad \mathbf{q}_2 = \frac{1}{\sqrt{2}} \begin{pmatrix} -1 \\ 1 \end{pmatrix} . \tag{6.4}$$

By direct calculation

$$\mathbf{q}_1{}^T \mathbf{q}_1 = \mathbf{q}_2{}^T \mathbf{q}_2 = 1 \tag{6.5}$$

(each column has unit length) and

$$\mathbf{q}_1{}^T\mathbf{q}_2 = \mathbf{q}_2{}^T\mathbf{q}_1 = 0 \tag{6.6}$$

(each column is orthogonal to every other column).

These column relations are true of every orthogonal matrix. And a non-orthogonal matrix fails to satisfy at least one of these relations.

Theorem 6.1. *A matrix is orthogonal if and only if each column has unit length and each column is orthogonal to every other column.*

Proof. The column relations are just the details of the matrix multiplication involved in $Q^TQ = I$. Namely, if $\mathbf{q}_i, \quad i = 1, \ldots, n$ are the columns of Q then

$$Q = \begin{pmatrix} \mathbf{q}_1 & \mathbf{q}_2 & \cdots & \mathbf{q}_n \end{pmatrix} \quad \text{and} \quad Q^T = \begin{pmatrix} \mathbf{q}_1{}^T \\ \mathbf{q}_2{}^T \\ \vdots \\ \mathbf{q}_n{}^T \end{pmatrix}$$

so

$$Q^TQ = \begin{pmatrix} \mathbf{q}_1{}^T\mathbf{q}_1 & \mathbf{q}_1{}^T\mathbf{q}_2 & \cdots & \mathbf{q}_1{}^T\mathbf{q}_n \\ \mathbf{q}_2{}^T\mathbf{q}_1 & \mathbf{q}_2{}^T\mathbf{q}_2 & \cdots & \mathbf{q}_2{}^T\mathbf{q}_n \\ \vdots & & & \vdots \\ \mathbf{q}_n{}^T\mathbf{q}_1 & \mathbf{q}_n{}^T\mathbf{q}_2 & \cdots & \mathbf{q}_n{}^T\mathbf{q}_n \end{pmatrix} = \begin{pmatrix} 1 & 0 & \cdots & 0 \\ 0 & 1 & \cdots & 0 \\ \vdots & & & \vdots \\ 0 & \cdots & 0 & 1 \end{pmatrix} . \quad \blacksquare$$

Example 6.1 Show that the matrix

$$R = R(\beta) = \begin{pmatrix} \cos\beta & -\sin\beta \\ \sin\beta & \cos\beta \end{pmatrix} \tag{6.7}$$

is an orthogonal matrix (for any number β) by

a) using Definition 6.1 and **b)** using Theorem 6.1.

Solution.

a) We calculate

$$RR^T = \begin{pmatrix} \cos\beta & -\sin\beta \\ \sin\beta & \cos\beta \end{pmatrix} \begin{pmatrix} \cos\beta & \sin\beta \\ -\sin\beta & \cos\beta \end{pmatrix} = \begin{pmatrix} 1 & 0 \\ 0 & 1 \end{pmatrix} = I$$

where we have used the identity $\cos^2\beta + \sin^2\beta = 1$. Similarly, $R^TR = I$, so by Definition 6.1, R is an orthogonal matrix for any value of β.

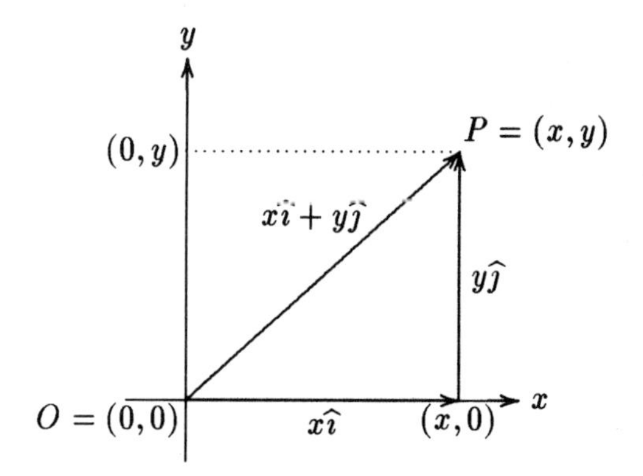

Figure 6.1

b) Let $\mathbf{q}_1 = \begin{pmatrix} \cos\beta \\ \sin\beta \end{pmatrix}$ and $\mathbf{q}_2 = \begin{pmatrix} -\sin\beta \\ \cos\beta \end{pmatrix}$ then

$$\mathbf{q}_1{}^T\mathbf{q}_1 = \begin{pmatrix} \cos\beta & \sin\beta \end{pmatrix} \begin{pmatrix} \cos\beta \\ \sin\beta \end{pmatrix} = 1$$

$$\mathbf{q}_1{}^T\mathbf{q}_2 = \begin{pmatrix} \cos\beta & \sin\beta \end{pmatrix} \begin{pmatrix} -\sin\beta \\ \cos\beta \end{pmatrix} = 0 \quad .$$

Similarly, $\mathbf{q}_2{}^T\mathbf{q}_2 = 1$ and $\mathbf{q}_2{}^T\mathbf{q}_1 = 0$. Thus R is an orthogonal matrix for any β.

These relations amongst the ***columns*** of an orthogonal matrix are equally true of the ***rows*** of an orthogonal matrix by the following argument. If Q is an orthogonal matrix, then Q^T is also an orthogonal matrix (*see exercise* 11), so the columns of Q^T satisfy the relations in Theorem 6.1. But the columns of Q^T are just the rows of Q.

6.2 Coordinates and Changes of Coordinates

A position vector $\overrightarrow{\mathbf{OP}}$ to a point P in the xy-plane $\mathfrak{R}^2$ is characterized by its ***Cartesian coordinates*** (x, y) with respect to a set of mutually perpendicular coordinate axes (*see Figure* 6.1).

If the unit position vector to $(1, 0)$ is denoted by $\hat{\imath} = \langle 1, 0\rangle$ and the unit position vector to $(0, 1)$ is denoted by $\hat{\jmath} = \langle 0, 1\rangle$, then $\overrightarrow{\mathbf{OP}} = \langle x, y\rangle = x\langle 1, 0\rangle + y\langle 0, 1\rangle = x\hat{\imath} + y\hat{\jmath}$. The directions of $\hat{\imath}$ and $\hat{\jmath}$ correspond to the positive directions along the coordinate axes. The expression $\overrightarrow{\mathbf{OP}} = x\hat{\imath} + y\hat{\jmath}$ uniquely represents an arbitrary point P of $\mathfrak{R}^2$ in terms of its coordinates (x, y) with respect to $\hat{\imath}$ and $\hat{\jmath}$. Consequently, $x\hat{\imath}+y\hat{\jmath} = \langle x, y\rangle = \langle 0,0\rangle = \mathbf{0}$ if and only if $x = y = 0$. This is an important property of $\{\hat{\imath}, \hat{\jmath}\}$. They form a ***linearly independent set of vectors.*** They form a ***basis*** or ***reference frame*** for $\mathcal{R}^2$. The set $\{\hat{\imath}, \hat{\jmath}\}$ is called the ***standard basis*** for $\mathcal{R}^2$. There are other bases for $\mathcal{R}^2$ and P has generally different coordinates with respect to them.

We need to make some definitions in order to investigate changes of coordinates.

Definition 6.2 In $\mathcal{R}^n$ if $\{\mathbf{a}_1, \ldots, \mathbf{a}_k\}$ is a set of vectors then a vector $\alpha_1\mathbf{a}_1+\cdots+\alpha_k\mathbf{a}_k$ is called a ***linear combination*** of $\{\mathbf{a}_1, \ldots, \mathbf{a}_k\}$ where each α_i $i = 1, \ldots, k$ is a real number.

Definition 6.3 In $\mathcal{R}^n$ if $\{\mathbf{a}_1, \ldots, \mathbf{a}_k\}$ is a set of vectors then they are ***linearly independent*** if the only linear combination which equals the zero vector $\mathbf{0}$ occurs when all the $\alpha_i = 0$, $i = 1, \ldots, k$, i.e.

$$\alpha_1\mathbf{a}_1 + \cdots + \alpha_k\mathbf{a}_k = \mathbf{0}$$

if and only if

$$\alpha_1 = \alpha_2 = \cdots = \alpha_k = 0 \quad .$$

Note that $\alpha_1'\mathbf{a}_1+\cdots+\alpha_k'\mathbf{a}_k = \alpha_1\mathbf{a}_1+\cdots+\alpha_k\mathbf{a}_k$ if and only if $(\alpha_1'-\alpha_1)\mathbf{a}_1+\cdots+(\alpha_k'-\alpha_k)\mathbf{a}_k = \mathbf{0}$. All of these coefficients must equal 0 if $\{\mathbf{a}_1, \ldots, \mathbf{a}_k\}$ is linearly independent or $\alpha_i' = \alpha_i$ for $i = 1, \ldots, k$. Thus linear combinations of linearly independent vectors are characterized by unique coefficients or ***coordinates*** $\{\alpha_1, \ldots, \alpha_k\}$ with respect to $\{\mathbf{a}_1, \ldots, \mathbf{a}_k\}$.

Definition 6.4 If $\{\mathbf{a}_1, \ldots, \mathbf{a}_k\}$ is a linearly independent set of k vectors in $\mathcal{R}^n$ then the set of all their linear combinations is a ***k-dimensional vector space*** V called the ***span of*** $\{\mathbf{a}_1, \ldots, \mathbf{a}_k\}$. A vector $\mathbf{b}$ is in V if and only if $\mathbf{b} = \alpha_1\mathbf{a}_1 + \cdots + \alpha_k\mathbf{a}_k$ where each α_i, $i = 1, \ldots, k$ is a real number. The set $\{\mathbf{a}_1, \ldots, \mathbf{a}_k\}$ is called a ***basis*** or ***reference frame*** for V. In particular if $k = n$ then $V = \mathcal{R}^n$ is an n-dimensional vector space. The ***standard basis*** for $\mathcal{R}^n$ is given by the n unit vectors $\langle 1, 0, \ldots, 0\rangle$, $\langle 0, 1, 0, \ldots, 0\rangle$, $\ldots$, $\langle 0, \ldots, 0, 1\rangle$ along the positive coordinate axes.

Note that any set of k linearly independent vectors in V is a basis for V. Thus, the dimension k of V equals the number of vectors in any basis for V.

The following example will illustrate some of these ideas in $\mathcal{R}^2$.

Example 6.2 Show that the vectors $\mathbf{a}_1 = \langle 3, 2\rangle$ and $\mathbf{a}_2 = \langle 1, 1\rangle$ are linearly independent in $\mathcal{R}^2$ and if P is a general point in the xy-plane express its position vector $\overrightarrow{\mathbf{OP}} = \langle x, y\rangle$ as a linear combination $\overrightarrow{\mathbf{OP}} = \alpha_1\mathbf{a}_1 + \alpha_2\mathbf{a}_2$.

Solution.

Set $\alpha_1\mathbf{a}_1 + \alpha_2\mathbf{a}_2 = \mathbf{0}$ then $\alpha_1\langle 3,2\rangle + \alpha_2\langle 1,1\rangle = \langle 0,0\rangle$ or

$$\begin{aligned} 3\alpha_1 + \alpha_2 &= 0 \\ 2\alpha_1 + \alpha_2 &= 0 \end{aligned}$$

or

$$\begin{pmatrix} 3 & 1 \\ 2 & 1 \end{pmatrix} \begin{pmatrix} \alpha_1 \\ \alpha_2 \end{pmatrix} = \begin{pmatrix} 0 \\ 0 \end{pmatrix} .$$

But

$$\begin{pmatrix} 3 & 1 \\ 2 & 1 \end{pmatrix}^{-1} = \begin{pmatrix} 1 & -1 \\ -2 & 3 \end{pmatrix}$$

so after multiplication by the inverse of the coefficient matrix it follows that

$$\begin{pmatrix} \alpha_1 \\ \alpha_2 \end{pmatrix} = \begin{pmatrix} 1 & -1 \\ -2 & 3 \end{pmatrix} \begin{pmatrix} 0 \\ 0 \end{pmatrix} = \begin{pmatrix} 0 \\ 0 \end{pmatrix}$$

or $\alpha_1 = \alpha_2 = 0$. Thus $\{\mathbf{a}_1, \mathbf{a}_2\}$ are linearly independent.

Set $\overrightarrow{\mathbf{OP}} = \alpha_1\mathbf{a}_1 + \alpha_2\mathbf{a}_2$ then

$$\alpha_1\langle 3,2\rangle + \alpha_2\langle 1,1\rangle = \langle x,y\rangle$$

or

$$\begin{aligned} 3\alpha_1 + \alpha_2 &= x \\ 2\alpha_1 + \alpha_2 &= y \ . \end{aligned}$$

Thus in matrix form we have

$$\begin{pmatrix} 3 & 1 \\ 2 & 1 \end{pmatrix} \begin{pmatrix} \alpha_1 \\ \alpha_2 \end{pmatrix} = \begin{pmatrix} x \\ y \end{pmatrix} \tag{6.8}$$

and after multiplication by the inverse of the coefficient matrix we have

$$\begin{pmatrix} \alpha_1 \\ \alpha_2 \end{pmatrix} = \begin{pmatrix} 1 & -1 \\ -2 & 3 \end{pmatrix} \begin{pmatrix} x \\ y \end{pmatrix} = \begin{pmatrix} x - y \\ -2x + 3y \end{pmatrix} . \tag{6.9}$$

We can represent a point $P = (x, y)$ in $\mathcal{R}^2$ by a column vector

$$\mathbf{v} = \begin{pmatrix} x \\ y \end{pmatrix} = [P] \tag{6.10}$$

its ***coordinate column with respect to the standard basis*** $\{\hat{\imath}, \hat{\jmath}\}$. In Example 6.2 we found that P had coordinates (α_1, α_2) with respect to $\mathbf{a} = \langle 3, 2 \rangle$ and $\mathbf{a}_2 = \langle 1, 1 \rangle$, that $(\alpha_1, \alpha_2) = (x - y, -2x + 3y)$, and that $(x, y) = (3\alpha_1 + \alpha_2, 2\alpha_1 + \alpha_2)$. Let

$$\mathbf{v}' = \begin{pmatrix} \alpha_1 \\ \alpha_2 \end{pmatrix} \tag{6.11}$$

be the ***coordinate column of*** P ***with respect to the new basis*** $\{\mathbf{a}_1, \mathbf{a}_2\}$. Then we have the following expression for the old coordinates in terms of the new ones

$$\mathbf{v} = A\mathbf{v}' \tag{6.12}$$

and the following expression for the new coordinates in terms of the old ones

$$\mathbf{v}' = A^{-1}\mathbf{v} \tag{6.13}$$

where

$$A = \begin{pmatrix} 3 & 1 \\ 2 & 1 \end{pmatrix} \tag{6.14}$$

and

$$A^{-1} = \begin{pmatrix} 1 & -1 \\ -2 & 3 \end{pmatrix} . \tag{6.15}$$

The matrix A in (6.13) is the ***matrix corresponding to the change of basis from*** $\{\hat{\imath}, \hat{\jmath}\}$ ***to*** $\{\mathbf{a}_1, \mathbf{a}_2\}$ or just simply the ***change of basis matrix***.

Note that

$$A \begin{pmatrix} 1 \\ 0 \end{pmatrix} = \begin{pmatrix} 3 & 1 \\ 2 & 1 \end{pmatrix} \begin{pmatrix} 1 \\ 0 \end{pmatrix} = \begin{pmatrix} 3 \\ 2 \end{pmatrix} = [\mathbf{a}_1]$$

and

$$A \begin{pmatrix} 0 \\ 1 \end{pmatrix} = \begin{pmatrix} 3 & 1 \\ 2 & 1 \end{pmatrix} \begin{pmatrix} 0 \\ 1 \end{pmatrix} = \begin{pmatrix} 1 \\ 1 \end{pmatrix} = [\mathbf{a}_2] .$$

Thus the coordinate columns of $\mathbf{a}_1$ and $\mathbf{a}_2$ with respect to $\{\hat{\imath}, \hat{\jmath}\}$ are the columns of the change of coordinate matrix A. Notice also that $\mathbf{a}_1 \bullet \mathbf{a}_2 = 5 \neq 0$, so $\mathbf{a}_1$ and $\mathbf{a}_2$ are not orthogonal. Thus the corresponding coordinate axes are not perpendicular. (See Figure 6.2.)

All of this extends to $\mathfrak{R}^n$ and we have the following:

Definition 6.5 If $\{\mathbf{a}_1, \ldots, \mathbf{a}_n\}$ is a basis for $\mathfrak{R}^n$ let A be the matrix whose j-th column $(j = 1, \ldots, n)$ is the coordinate column $[\mathbf{a}_j]$ with respect to the standard basis for $\mathfrak{R}^n$. Then A is the ***matrix corresponding to the change of basis from the standard basis to*** $\{\mathbf{a}_1, \ldots, \mathbf{a}_n\}$ or simply the ***change of basis matrix.***

Note that if $\mathbf{v}$ is the coordinate column of a point P with respect to the standard basis and if $\mathbf{v}'$ is the coordinate column of P with respect to the new basis $\{\mathbf{a}_1, \ldots, \mathbf{a}_n\}$ then

$$\mathbf{v} = A\mathbf{v}' \tag{6.16}$$

expresses the old standard coordinates of P in terms of its new coordinates and

$$\mathbf{v}' = A^{-1}\mathbf{v} \tag{6.17}$$

expresses the new coordinates of P in terms of its old standard coordinates.

If the new basis vectors $\{\mathbf{a}_1, \ldots, \mathbf{a}_n\}$ are not unit orthogonal vectors the change of basis matrix A need not necessarily be an orthogonal matrix, so in general it is necessary to use the standard methods of Chapter 5 to compute A^{-1} and find $\mathbf{v}$.

As we have already noted at the end of Section 6.1, it is much easier and convenient to find the inverse of an orthogonal matrix. In the remainder of this chapter we will consider changes of basis where the matrix A is an orthogonal matrix Q.

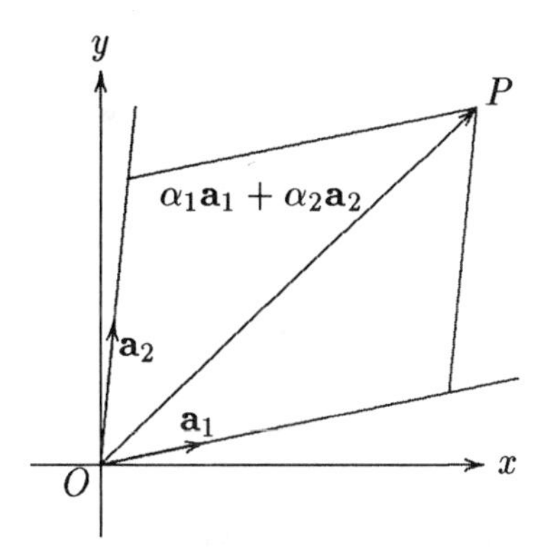

Figure 6.2

6.3 Changes of Coordinates with Orthogonal Matrices

In order to change bases in $\mathcal{R}^n$ we need n linearly independent vectors. Suppose we have k mutually orthogonal unit vectors (an ***orthonormal set of vectors***) $\{\mathbf{u}_1, \ldots, \mathbf{u}_k\}$. This means each $\mathbf{u}_i$ has length 1 or $\mathbf{u}_i \bullet \mathbf{u}_i = 1$ and that for $i \neq j$ $\mathbf{u}_i \bullet \mathbf{u}_j = 0$. Then the following result shows that they must form a linearly independent set of vectors.

Theorem 6.2. *Any orthonormal set of k vectors $\{\mathbf{u}_1, \ldots, \mathbf{u}_k\}$ in $\mathcal{R}^n$ must be linearly independent and so any orthonormal set of n vectors in $\mathcal{R}^n$ must be a basis for $\mathcal{R}^n$.*

Proof. Consider a zero linear combination $\mathbf{0} = \alpha_1 \mathbf{u}_1 + \ldots + \alpha_k \mathbf{u}_k$. Take the dot product on both sides of the equation with respect to $\mathbf{u}_1, \ldots, \mathbf{u}_k$ to obtain

$$\begin{array}{rcccl} 0 & = & \alpha_1 \mathbf{u}_1 \bullet \mathbf{u}_1 + 0 + \cdots + 0 & = & \alpha_1 \\ 0 & = & 0 + \alpha_2 \mathbf{u}_2 \bullet \mathbf{u}_2 + \ldots + 0 & = & \alpha_2 \\ & & \vdots & & \\ 0 & = & 0 + \cdots + 0 + \alpha_k \mathbf{u}_k \bullet \mathbf{u}_k & = & \alpha_k \end{array} .$$

Thus only $\alpha_1 = \cdots = \alpha_k = 0$ is possible, so $\{\mathbf{u}_1, \ldots, \mathbf{u}_k\}$ is linearly independent. ∎

Let us investigate changes of basis in $\mathcal{R}^2$ with an orthogonal change of basis matrix Q.

Example 6.3 Determine the form of an arbitrary 2×2 orthogonal matrix Q and consequently determine the possible corresponding new basis vectors for $\mathcal{R}^2$.

Solution.

Let Q be an arbitrary orthogonal matrix

$$Q = \begin{pmatrix} q_{11} & q_{12} \\ q_{21} & q_{22} \end{pmatrix} .$$

By its definition we know that $Q^T = Q$ and, therefore, $Q^T Q = QQ^T = I$ by the definition of an inverse. Then $QQ^T = I$ gives us

$$\begin{pmatrix} q_{11} & q_{12} \\ q_{21} & q_{22} \end{pmatrix} \begin{pmatrix} q_{11} & q_{21} \\ q_{12} & q_{22} \end{pmatrix} =$$

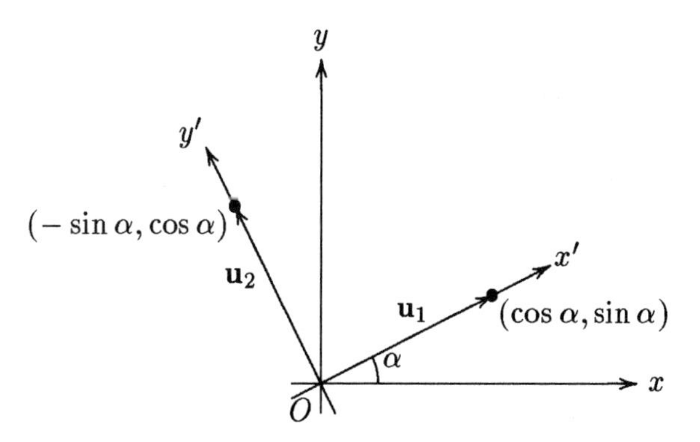

Figure 6.3

$$\begin{pmatrix} q_{11}^2 + q_{12}^2 & q_{11}q_{21} + q_{12}q_{22} \\ q_{21}q_{11} + q_{22}q_{12} & q_{21}^2 + q_{22}^2 \end{pmatrix} = \begin{pmatrix} 1 & 0 \\ 0 & 1 \end{pmatrix}$$

Hence, writing out the distinct equations, we obtain

$$\begin{aligned} q_{11}^2 + q_{12}^2 &= 1 \\ q_{21}^2 + q_{22}^2 &= 1 \\ q_{21}q_{11} + q_{22}q_{12} &= 0 \ . \end{aligned} \tag{6.18}$$

From the first two equations we see that (q_{11}, q_{12}) and (q_{21}, q_{22}) are points on the unit circle $x^2 + y^2 = 1$, so there exists angles α and β such that $(q_{11}, q_{12}) = (\cos\alpha, -\sin\alpha)$ and $(q_{21}, q_{22}) = (\sin\beta, \cos\beta)$. Now we can rewrite the last equation

$$\sin\beta\cos\alpha - \cos\beta\sin\alpha = \sin(\beta - \alpha) = 0$$

by a trigonometric identity. Solving for β in terms of α we obtain $\beta = \alpha$ or $\beta = \alpha + \pi$ up to a multiple of 2π. Substitution of these values into $(q_{21}, q_{22}) = (\sin\beta, \cos\beta)$ we obtain $(q_{21}, q_{22}) = (\sin\alpha, \cos\alpha)$ or $(-\sin\alpha, -\cos\alpha)$.

In the first case Q is given by

$$Q = \begin{pmatrix} \cos\alpha & -\sin\alpha \\ \sin\alpha & \cos\alpha \end{pmatrix} \ . \tag{6.19}$$

Consequently, the vectors $\mathbf{u}_1 = \langle\cos\alpha, \sin\alpha\rangle$ and $\mathbf{u}_2 = \langle-\sin\alpha, \cos\alpha\rangle$ determine the directions of the new coordinate axes.

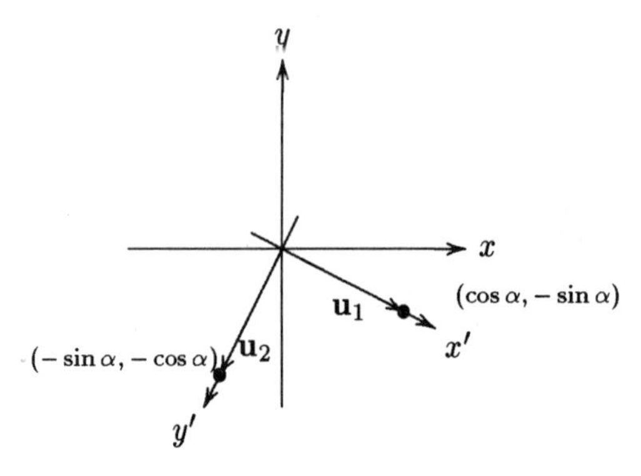

Figure 6.4

As we can see this corresponds to a rotation of coordinate axes by an angle α about the origin. (*Figure 6.3*) The matrix Q in (6.19) equals the matrix $R = R(\alpha)$ from Example 6.1 and is called a ***rotation matrix***.

In the second case Q is given by

$$Q = \begin{pmatrix} \cos\alpha & -\sin\alpha \\ -\sin\alpha & -\cos\alpha \end{pmatrix} . \tag{6.20}$$

Consequently, the vectors $\mathbf{u}_1 = \langle \cos\alpha, -\sin\alpha \rangle$ and $\mathbf{u}_2 = \langle -\sin\alpha, -\cos\alpha \rangle$ determine the directions of the new coordinate axes.

As we can see this corresponds to a reflection of the y-axis through the origin followed by a rotation of the resulting axes by the angle $-\alpha$.

Note that for $Q = R(\alpha)$ from (6.19) we have $\det R = \cos^2\alpha + \sin^2\alpha = 1$ and that for the Q from (6.20) we have $\det Q = -\cos^2\alpha - \sin^2\alpha = -1$. Thus the rotation matrix $R(\alpha)$ is an orthogonal matrix with determinant 1. This property is characteristic of matrices corresponding to rotations of coordinate axes. Also note that under a rotation of coordinate axes the ***orientation*** of the basis vectors is unchanged. However, if a reflection and rotation of the coordinate axes occurs then the orientation is reversed. (*Figure 6.4*)

Definition 6.6 An $n \times n$ orthogonal matrix Q with $\det Q = 1$ is called a ***rotation matrix*** and corresponds to a rotation of coordinate axes. The columns of $Q = [\mathbf{q}_1, \ldots, \mathbf{q}_n]$ have the same ***orientation*** (*relative positions*) as the columns of the $n \times n$ identity matrix $I = [\mathbf{e}_1, \ldots, \mathbf{e}_n]$. An orthogonal matrix Q with $\det Q = -1$ thus corresponds to a ***rotation and a reflection*** of coordinate axes. In this case the columns of $Q = [\mathbf{q}_1, \ldots, \mathbf{q}_n]$ have the ***opposite orientation*** to that of the columns of $I = [\mathbf{e}_1, \ldots, \mathbf{e}_n]$. For $n = 3$ the columns of a rotation matrix are a ***right handed set of vectors*** and for Q with $\det Q = -1$ its columns are a ***left handed set of vectors***.

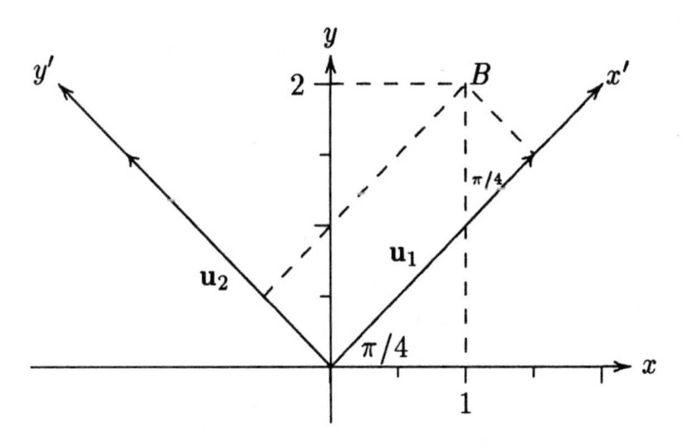

Figure 6.5

Generally, we will want to keep the orientation of our coordinate axes unchanged and so will use rotation matrices to change coordinates in Chapter 8.

If $Q = [\mathbf{q}_1, \ldots, \mathbf{q}_n]$ is any $n \times n$ orthogonal matrix with $\det Q = -1$ then $Q' = [-\mathbf{q}_1, \mathbf{q}_2, \ldots, \mathbf{q}_n]$ has $\det Q' = 1$ and corresponds to the same set of coordinate lines through the origin as Q. Only now Q' is a rotation matrix. Thus, once the coordinate lines (the *unoriented coordinate axes*) are determined it is easy to find a rotation matrix which rotates the old coordinate axes onto the coordinate lines.

Let us now look at some examples in $\mathfrak{R}^2$ with specific angles of rotation.

Example 6.4 Suppose the coordinate axes are rotated by an angle $\alpha = \frac{\pi}{4}$ (see Figure 6.5)

The change of coordinate matrix Q is given by

$$Q = \frac{1}{\sqrt{2}} \begin{pmatrix} 1 & -1 \\ 1 & 1 \end{pmatrix} . \tag{6.21}$$

The new basis vectors are $\mathbf{u}_1 = \langle \frac{1}{\sqrt{2}}, \frac{1}{\sqrt{2}} \rangle$ and $\mathbf{u}_2 = \langle \frac{-1}{\sqrt{2}}, \frac{1}{\sqrt{2}} \rangle$. The old coordinates (x, y) of a point P are given in terms of the new primed coordinates (x', y') by $\mathbf{v} = Q\mathbf{v}'$ or

$$\begin{pmatrix} x \\ y \end{pmatrix} = \frac{1}{\sqrt{2}} \begin{pmatrix} 1 & -1 \\ 1 & 1 \end{pmatrix} \begin{pmatrix} x' \\ y' \end{pmatrix} = \frac{1}{\sqrt{2}} \begin{pmatrix} x' - y' \\ x' + y' \end{pmatrix} . \tag{6.22}$$

The new primed coordinates (x', y') of P are given in terms of the old coordinates (x, y) by $\mathbf{v}' = Q^T\mathbf{v}$ or

$$\begin{pmatrix} x' \\ y' \end{pmatrix} = \frac{1}{\sqrt{2}} \begin{pmatrix} 1 & 1 \\ -1 & 1 \end{pmatrix} \begin{pmatrix} x \\ y \end{pmatrix} = \frac{1}{\sqrt{2}} \begin{pmatrix} x + y \\ -x + y \end{pmatrix} . \tag{6.23}$$

Using (6.23) if B is a point with old unprimed coordinates $(x, y) = (1, 2)$ with respect to the old coordinate axes then its new primed coordinates (x', y') with respect to the new coordinate axes are given by

$$\begin{pmatrix} x' \\ y' \end{pmatrix} = \frac{1}{\sqrt{2}} \begin{pmatrix} 1 & 1 \\ -1 & 1 \end{pmatrix} \begin{pmatrix} 1 \\ 2 \end{pmatrix} = \frac{1}{\sqrt{2}} \begin{pmatrix} 3 \\ 1 \end{pmatrix} .$$

Using (6.22) if C is a point with new primed coordinates $(x', y') = (2, 1)$ then its old unprimed coordinates (x, y) are given by

$$\begin{pmatrix} x \\ y \end{pmatrix} = \frac{1}{\sqrt{2}} \begin{pmatrix} 1 & -1 \\ 1 & 1 \end{pmatrix} \begin{pmatrix} 2 \\ 1 \end{pmatrix} = \frac{1}{\sqrt{2}} \begin{pmatrix} 1 \\ 3 \end{pmatrix} .$$

Example 6.5 Find the orthogonal matrix corresponding to a rotation of coordinate axes by an angle $\alpha = \frac{\pi}{2}$ and find the new basis vectors.

Solution.

$$Q = R(\tfrac{\pi}{2}) = \begin{pmatrix} 0 & -1 \\ 1 & 0 \end{pmatrix}$$

so $\mathbf{u}_1 = \langle 0, 1 \rangle = \hat{\jmath}$ and $\mathbf{u}_2 = \langle -1, 0 \rangle = -\hat{\imath}$.

Example 6.6 Show that

$$Q = \begin{pmatrix} -1 & 0 \\ 0 & 1 \end{pmatrix}$$

is an orthogonal matrix. Then find the new basis vectors. If A and B are points with old coordinates (α_1, α_2) and (β_1, β_2) find their new primed coordinates (α_1', α_2') and (β_1', β_2'). Then $\mathbf{a} = [\alpha_1, \alpha_2]^T$, $\mathbf{b} = [\beta_1, \beta_2]^T$, $\mathbf{a}' = [\alpha_1', \alpha_2']^T$ and $b' = [\beta_1', \beta_2']^T$. Show that $\mathbf{a}^T\mathbf{b} = \mathbf{a}'^T\mathbf{b}'$, i.e. dot products do not change values if new coordinates are introduced with an orthogonal matrix.

Solution.

We have $Q^T = Q$ and $Q^2 = I$ so $Q^T = Q^{-1} = Q$ and is an orthogonal matrix. Then $\mathbf{u}_1 = \langle -1, 0 \rangle = -\hat{\imath}$ and $\mathbf{u}_2 = \langle 0, 1 \rangle = \hat{\jmath}$.

Now from $\mathbf{a}' = Q^T\mathbf{a} = Q\mathbf{a}$ and $\mathbf{b}' = Q^T\mathbf{b} = Q\mathbf{b}$ we have

$$\begin{pmatrix} \alpha_1' \\ \alpha_2' \end{pmatrix} = \begin{pmatrix} -1 & 0 \\ 0 & 1 \end{pmatrix} \begin{pmatrix} \alpha_1 \\ \alpha_2 \end{pmatrix} = \begin{pmatrix} -\alpha_1 \\ \alpha_2 \end{pmatrix}$$

and

$$\begin{pmatrix} \beta_1' \\ \beta_2' \end{pmatrix} = \begin{pmatrix} -1 & 0 \\ 0 & 1 \end{pmatrix} \begin{pmatrix} \beta_1 \\ \beta_2 \end{pmatrix} = \begin{pmatrix} -\beta_1 \\ \beta_2 \end{pmatrix} .$$

Therefore, $\mathbf{a}^T\mathbf{b} = \langle \alpha_1, \alpha_2 \rangle \bullet \langle \beta_1, \beta_2 \rangle = \alpha_1\beta_1 + \alpha_2\beta_2 = \langle -\alpha_1, \alpha_2 \rangle \bullet \langle -\beta_1, \beta_2 \rangle = \mathbf{a}'^T\mathbf{b}'$.

In general if Q is any orthogonal matrix and if two points have coordinate columns $\mathbf{v}$ and $\mathbf{w}$ with respect to the old standard basis and coordinate columns $\mathbf{v}'$ and $\mathbf{w}'$ with respect to the new basis then $\mathbf{v} = Q\mathbf{v}'$ and $\mathbf{w} = Q\mathbf{w}'$. The dot product of $\mathbf{v}$ and $\mathbf{w}$ is given by $\mathbf{v} \bullet \mathbf{w} = \mathbf{v}^T\mathbf{w} = (Q\mathbf{v}')^T Q\mathbf{w}' = \mathbf{v}'Q^TQ\mathbf{w}' = \mathbf{v}'^T\mathbf{w}' = \mathbf{v}' \bullet \mathbf{w}'$ since $Q^T = Q^{-1}$. Thus, dot products are independent of the orthogonal coordinate system used and can be computed using coordinates with respect to any convenient orthonormal basis.

We will finish our discussion of coordinate changes with orthogonal matrices by looking at the three dimensional case.

Suppose the columns $\{\mathbf{q}_1, \mathbf{q}_2, \mathbf{q}_3\}$ of a 3×3 rotation matrix $Q = [\mathbf{q}_1, \mathbf{q}_2, \mathbf{q}_3]$ form a right handed set of orthonormal vectors. Thus, $\mathbf{q}_i \bullet \mathbf{q}_j = 0$ if $i \neq j$, each $\mathbf{q}_i$ has length 1, $\mathbf{q}_1 \times \mathbf{q}_2 = \mathbf{q}_3$, $\mathbf{q}_2 \times \mathbf{q}_3 = \mathbf{q}_1$, and $\mathbf{q}_3 \times \mathbf{q}_1 = \mathbf{q}_2$ (*where* $\times$ *denotes the vector cross product in* $\mathfrak{R}^3$). The determinant of Q equals $\mathbf{q}_1 \times \mathbf{q}_2 \bullet \mathbf{q}_3 = q$.

Example 6.7 Find numbers $a_1 > 0$, a_2, b_1, b_2, and b_3 so that

$$Q = \begin{pmatrix} \frac{1}{\sqrt{3}} & a_1 & b_1 \\ \frac{1}{\sqrt{3}} & a_2 & b_2 \\ \frac{-1}{\sqrt{3}} & 3a_1 & b_3 \end{pmatrix}$$

is a rotation matrix.

Solution.

From

$$\mathbf{q}_1 = \frac{1}{\sqrt{3}} \begin{pmatrix} 1 \\ 1 \\ -1 \end{pmatrix}, \qquad \mathbf{q}_2 = \begin{pmatrix} a_1 \\ a_2 \\ 3a_1 \end{pmatrix}, \qquad \mathbf{q}_3 = \begin{pmatrix} b_1 \\ b_2 \\ b_3 \end{pmatrix}$$

we have $\mathbf{q}_1^T\mathbf{q}_2 = 0$ and $\mathbf{q}_1^T\mathbf{q}_3 = 0$. Hence we obtain the system

$$\begin{aligned} a_1 + a_2 - 3a_1 &= 0 \\ b_1 + b_2 - b_3 &= 0 \end{aligned} .$$

From the first equation we find that $a_2 - 2a_1 = 0$ or $a_2 = 2a_1$. Thus $\mathbf{q}_2$ is given by

$$\mathbf{q}_2 = a_1 \begin{pmatrix} 1 \\ 2 \\ 3 \end{pmatrix} .$$

Using $|\mathbf{q}_2| = 1$ we obtain $1 = |\mathbf{a}_1|\sqrt{1+4+9} = .|\mathbf{a}_1|$. Thus $a_1 = \frac{1}{\sqrt{14}}$ then

$$\mathbf{q}_2 = \frac{1}{\sqrt{14}} \begin{pmatrix} 1 \\ 2 \\ 3 \end{pmatrix} .$$

Now, $\mathbf{q}_3$ is completely determined by $\mathbf{q}_3 = \mathbf{q}_1 \times \mathbf{q}_2$ and we obtain

$$\mathbf{q}_3 = \frac{1}{\sqrt{42}} \begin{pmatrix} 5 \\ -4 \\ 1 \end{pmatrix} .$$

Thus, Q is given by

$$Q = \begin{pmatrix} \frac{1}{\sqrt{3}} & \frac{1}{\sqrt{14}} & \frac{5}{\sqrt{42}} \\ \frac{1}{\sqrt{3}} & \frac{2}{\sqrt{14}} & \frac{-4}{\sqrt{42}} \\ \frac{-1}{\sqrt{3}} & \frac{3}{\sqrt{14}} & \frac{1}{\sqrt{42}} \end{pmatrix} .$$

Example 6.8 Show that

$$Q = \begin{pmatrix} \frac{-1}{\sqrt{2}} & \frac{1}{\sqrt{3}} & \frac{1}{\sqrt{6}} \\ \frac{1}{\sqrt{2}} & \frac{1}{\sqrt{3}} & \frac{1}{\sqrt{6}} \\ 0 & \frac{1}{\sqrt{3}} & \frac{-2}{\sqrt{6}} \end{pmatrix} \tag{6.24}$$

is an orthogonal matrix, determine whether or not it is a rotation matrix, and find new orthonormal basis vectors.

Solution.

The columns of Q are given by

$$\mathbf{q}_1 = \frac{1}{\sqrt{2}} \begin{pmatrix} -1 \\ 1 \\ 0 \end{pmatrix}, \quad \mathbf{q}_2 = \frac{1}{\sqrt{3}} \begin{pmatrix} 1 \\ 1 \\ 1 \end{pmatrix}, \quad \text{and} \quad \mathbf{q}_3 = \frac{1}{\sqrt{6}} \begin{pmatrix} 1 \\ 1 \\ -2 \end{pmatrix} .$$

For this matrix, the easiest way to show that Q is orthogonal is to show that $\mathbf{q}_1^T\mathbf{q}_2 = \mathbf{q}_1 \bullet \mathbf{q}_2 = 0$, $\mathbf{q}_1^T\mathbf{q}_3 = \mathbf{q}_1 \bullet \mathbf{q}_3 = 0$, and that $|\mathbf{q}_1| = |\mathbf{q}_2| = |\mathbf{q}_3| = 1$. This can easily be done so we see that Q is an orthogonal matrix. To check whether or not Q is a rotation matrix compute $\det Q$. Pulling out the common factors from each column of Q we obtain

$$\det Q = \frac{1}{\sqrt{2}}\frac{1}{\sqrt{3}}\frac{1}{\sqrt{6}}\begin{vmatrix} -1 & 1 & 1 \\ 1 & 1 & 1 \\ 0 & 1 & -2 \end{vmatrix} = \frac{1}{6}\begin{vmatrix} 0 & 2 & 2 \\ 1 & 1 & 1 \\ 0 & 1 & -2 \end{vmatrix} = \frac{-1}{6}\begin{vmatrix} 2 & 2 \\ 1 & -2 \end{vmatrix} = \frac{-1}{6}(-4 - 2) = 1$$

so Q is a rotation matrix. The new orthonormal basis vectors are given by $\mathbf{u}_1 = \frac{1}{\sqrt{2}}\langle -1, 1, 0\rangle$, $\mathbf{u}_2 = \frac{1}{\sqrt{3}}\langle 1, 1, 1\rangle$, and $\mathbf{u}_3 = \frac{1}{\sqrt{6}}\langle 1, 1, -2\rangle$.

Example 6.9 The orthogonal matrix Q' is obtained form the orthogonal matrix Q in Example 6.8 by the interchange of the first two columns, i.e. $Q' = [\mathbf{q}_2, \mathbf{q}_1, \mathbf{q}_3] = [\mathbf{q'}_1, \mathbf{q'}_2, \mathbf{q'}_3]$ or

$$Q' = \begin{pmatrix} \frac{1}{\sqrt{3}} & \frac{-1}{\sqrt{2}} & \frac{1}{\sqrt{6}} \\ \frac{1}{\sqrt{3}} & \frac{1}{\sqrt{2}} & \frac{1}{\sqrt{6}} \\ \frac{1}{\sqrt{3}} & 0 & \frac{-2}{\sqrt{6}} \end{pmatrix} .$$

Show that Q' is not a rotation matrix and find the new orthonormal basis vectors.

Solution.

Since $\det Q' = -\det Q = -1$ from the properties of determinants we see that Q' is not a rotation matrix. The new orthonormal basis vectors are given by $\mathbf{u}_1 = \frac{1}{\sqrt{3}}\langle 1, 1, 1\rangle$, $\mathbf{u}_2 = \frac{1}{\sqrt{2}}\langle -1, 1, 0\rangle$, and $\mathbf{u}_3 = \frac{1}{\sqrt{6}}\langle 1, 1, -2\rangle$. Observe that $\mathbf{u}_1 \times \mathbf{u}_2 = -\mathbf{u}_3$ so $\{\mathbf{u}_1, \mathbf{u}_2, \mathbf{u}_3\}$ is a left handed set of vectors.

6.4 Exercises

1. Verify whether or not the following matrices are orthogonal.

a) $\begin{pmatrix} 1 & -1 \\ 0 & 1 \end{pmatrix}$ **b)** $\begin{pmatrix} \frac{\sqrt{3}}{2} & \frac{-1}{2} \\ \frac{1}{2} & \frac{\sqrt{3}}{2} \end{pmatrix}$ **c)** $\begin{pmatrix} 1 & 0 & 0 \\ 0 & 0 & 1 \\ 0 & 1 & 0 \end{pmatrix}$

d) $\begin{pmatrix} \frac{1}{\sqrt{3}} & \frac{2}{\sqrt{6}} & 0 \\ \frac{1}{\sqrt{3}} & \frac{-1}{\sqrt{6}} & \frac{1}{\sqrt{2}} \\ \frac{1}{\sqrt{3}} & \frac{-1}{\sqrt{6}} & \frac{-1}{\sqrt{2}} \end{pmatrix}$ **e)** $\begin{pmatrix} \cos\beta & -\sin\beta & 0 \\ \sin\beta & \cos\beta & 0 \\ 0 & 0 & 1 \end{pmatrix}$ **f)** $\begin{pmatrix} 2 & 1 & 2 \\ 1 & 2 & -2 \\ -2 & 2 & 1 \end{pmatrix}$.

2. Let $Q = \frac{1}{3}\begin{pmatrix} 1 & 2 & -2 \\ -2 & 2 & 1 \\ 2 & 1 & 2 \end{pmatrix}$ be a change of basis matrix. It corresponds to a rotation of coordinate axes. Let $\mathbf{v} = [x, y, z]^T$ and $\mathbf{v}' = [x', y', z']^T$ be the coordinate columns of a point with respect to old standard basis and the new rotated basis respectively.

a) Find the old (standard) coordinates of following points with new primed coordinates given by $(1,0,0)$, $(0,1,0)$, and $(0,0,1)$.

b) Find the new primed coordinates of the point with old unprimed coordinates $(6,0,3)$.

c) Find the old unprimed coordinates of the point with new primed coordinates $(6,0,3)$.

3. Repeat exercise 2 but with $Q = \begin{pmatrix} \frac{1}{\sqrt{2}} & \frac{-1}{\sqrt{2}} & 0 \\ \frac{1}{\sqrt{2}} & \frac{1}{\sqrt{2}} & 0 \\ 0 & 0 & 1 \end{pmatrix}$ and with coordinates $(1,1,1)$ in parts **b)** and **c)**.

4. Is there a number r such that $Q = \begin{pmatrix} 0 & 1 \\ r & 3 \end{pmatrix}$ is orthogonal?

5. Find numbers a_1, b_1, c_1, b_2, c_2, and c_3 so that $Q = \begin{pmatrix} a_1 & b_1 & c_1 \\ 0 & b_2 & c_2 \\ 0 & 0 & c_3 \end{pmatrix}$ is orthogonal.

6. Find numbers a_1, a_2, b_1, b_2, and b_3 so that $Q = [\mathbf{q}_1, \mathbf{q}_2, \mathbf{q}_3]$ is a rotation matrix where

$$\mathbf{q}_1 = \frac{1}{\sqrt{6}}\begin{pmatrix} 1 \\ 2 \\ -1 \end{pmatrix}, \qquad \mathbf{q}_2 = \begin{pmatrix} a_1 \\ a_2 \\ 3a_1 \end{pmatrix}, \qquad \mathbf{q}_3 = \begin{pmatrix} b_1 \\ b_2 \\ b_3 \end{pmatrix}.$$

7. The matrix

$$\begin{pmatrix} \frac{1}{2} & \frac{-\sqrt{3}}{2} \\ \frac{\sqrt{3}}{2} & \frac{1}{2} \end{pmatrix}$$

corresponds to a rotation of coordinate axes through an acute angle β.

a) Find β.

b) Sketch the coordinate axes in the original and rotated systems and find the new orthonormal basis.

8. Repeat Problem 7 for $\begin{pmatrix} \frac{1}{\sqrt{2}} & \frac{-1}{\sqrt{2}} \\ \frac{1}{\sqrt{2}} & \frac{1}{\sqrt{2}} \end{pmatrix}$.

9. What is the change of basis matrix corresponding to the change of coordinates such that the x'-axis is the y-axis, the y'-axis is the negative of the z-axis, and the z'-axis is the x-axis? Is this a rotation of coordinate axes?

10. The matrix $\begin{pmatrix} \frac{-1}{\sqrt{2}} & \frac{1}{\sqrt{2}} \\ \frac{1}{\sqrt{2}} & \frac{1}{\sqrt{2}} \end{pmatrix}$ is an orthogonal change of basis matrix. Determine whether or not it is a rotation matrix. Sketch the pairs of coordinate axes.

11. Prove that if Q is an orthogonal matrix if and only if Q^T is an orthogonal matrix.

12. Show that the determinant of an orthogonal matrix Q can only be $+1$ or -1.

13. Let $Q = \frac{1}{7}\begin{pmatrix} -6 & 3 & 2 \\ 3 & 2 & 6 \\ -2 & -6 & 3 \end{pmatrix}$ be the change of basis matrix from the old (standard) basis to a new orthonormal basis. Convert the following equation in the old (x, y, z) coordinates

$$x - y - z = 6$$

into an equation involving the new (x', y', z') coordinates.

14. Show that the product of two orthogonal matrices is orthogonal.

15. Assume that I is the 3×3 identity matrix and $\mathbf{a}$ is a non-zero column vector with 3 components. Show that

$$I - \frac{2}{|\mathbf{a}|^2}\mathbf{a}\mathbf{a}^T$$

is an orthogonal matrix.

16. Let $Q = \frac{1}{7}\begin{pmatrix} 6 & 3 & -2 \\ -3 & 2 & -6 \\ -2 & 6 & 3 \end{pmatrix}$ be the change of basis matrix from the old (standard) basis to a new orthonormal basis. Convert the following from the old (x, y, z) coordinates to the new (x', y', z') coordinates.

a) $(6, 2, 1)$ **b)** $(7, 3, -1)$ **c)** $(11, 3, 4)$.

17. Let $Q = \frac{1}{7}\begin{pmatrix} 3 & 2 & 6 \\ -2 & 6 & -3 \\ 6 & 3 & 2 \end{pmatrix}$ be the change of basis matrix from the old (standard) basis to a new orthonormal basis. Convert the following from the new (x', y', z') coordinates to the old (x, y, z) coordinates.

a) $(4,3,1)$ **b)** $(-4,1,7)$ **c)** $(-4,3,11)$.

18. Let $Q = \frac{1}{9}\begin{pmatrix} 8 & -4 & 1 \\ 4 & 7 & -4 \\ 1 & 4 & 8 \end{pmatrix}$ be the change of basis matrix from the old (standard) basis to a new orthonormal basis. Convert the following from the old (x, y, z) coordinates to the new (x', y', z') coordinates.

a) $(-8,5,-1)$ **b)** $(-5,1,0)$ **c)** $(4,-9,4)$.

7 The Eigenvalue Problem

Outline of Key Ideas

Features of the eigenvalue problem:

a geometric example

definition of eigenvalues and eigenvectors.

Calculating eigenvalues and eigenvectors of a matrix:

eigenvalues are solutions of the characteristic equation

$$\det(A - \lambda I) = 0$$

eigenvectors are non-zero solutions of the homogeneous linear systems

$$(A - \lambda I)\mathbf{v} = 0 \quad .$$

The existence of a complete set of mutually orthogonal eigenvectors:

a symmetric matrix has a complete set of mutually orthogonal eigenvectors.

a technique for finding a complete set.

The Eigenvalue Problem

7.1 Features of the Eigenvalue Problem

The eigenvalue problem occurs in many areas of applied mathematics. For example mechanical systems involving several masses and springs can be treated as eigenvalue problems. Another example is the study of vibrating elastic membranes which are stretched over fixed circular shaped boundaries. The principal directions are given by eigenvectors. In the theory of elasticity the principal directions of stress can strain are given by the eigenvectors of the stress-strain tensor which is a symmetric matrix. Finally, eigenvalues occur in the solutions to the set of differential equations describing the dynamics (*movement characteristics*) of a suspension bridge. The square roots of these eigenvalues are the natural frequencies of oscillation of the bridge. Knowledge of these natural frequencies of oscillation of structures can be very useful. For example, if wind gusts beat against a bridge with a frequency matching one of its natural frequencies, the bridge will begin to oscillate in resonance with the wind gusts. The amplitude of the oscillation may increase to the point where the structure collapses. The Tacoma Narrows Bridge appears to have executed just such oscillations before collapsing into Puget Sound in 1940. Another example involves a playground swing. If the person on the swing adds energy at a frequency corresponding to the natural frequency of the swing, then the swing goes higher.

In order to introduce the eigenvalue problem let us use a simple geometric model. In the xy-plane, i.e. $\Re^2$, let us find the point of reflection Q of a general point $P = (a, b)$ in the line $y = x$.

The line $x - y = 0$ has a normal vector $\langle 1, -1 \rangle$ and so a vector equation of the straight line through P and perpendicular to $y = x$ is given by $\langle x, y \rangle = t\langle 1, -1 \rangle + \langle a, b \rangle = \langle a + t, b - t \rangle$. The point of intersection $M = (x, y)$ of the two lines is obtained from $a + t = b - t$

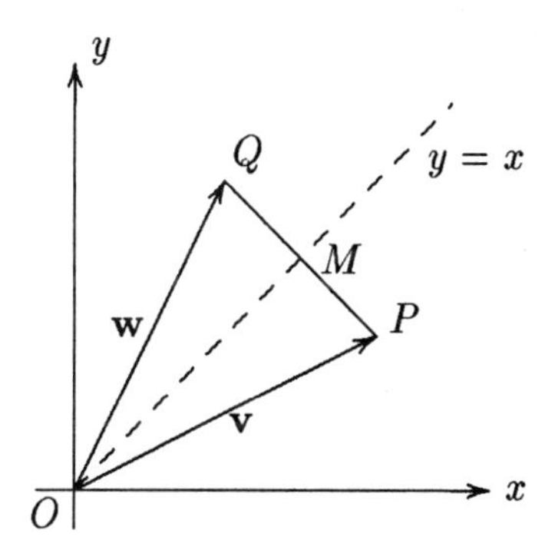

Figure 7.1

or $t = \frac{1}{2}(b-a)$. Thus, $M = (\frac{1}{2}(a+b), \frac{1}{2}(a+b))$. Now the point Q is obtained from $\overrightarrow{\mathbf{OQ}} = \overrightarrow{\mathbf{OP}} + 2\overrightarrow{\mathbf{PM}}$ or $\overrightarrow{\mathbf{OQ}} = \langle a,b\rangle + 2\langle \frac{1}{2}(b-a), \frac{1}{2}(a-b)\rangle = \langle a+(b-a), b+(a-b)\rangle = \langle b,a\rangle$. Thus $Q=(b,a)$ when $P=(a,b)$. In matrix form if $\mathbf{w}=[Q]$ and $\mathbf{v}=[P]$ then

$$\begin{pmatrix} b \\ a \end{pmatrix} = \begin{pmatrix} 0 & 1 \\ 1 & 0 \end{pmatrix} \begin{pmatrix} a \\ b \end{pmatrix}$$

or

$$\mathbf{w} = A\mathbf{v}$$

where A is the matrix

$$A = \begin{pmatrix} 0 & 1 \\ 1 & 0 \end{pmatrix} .$$

In this case the matrix product $A\mathbf{v}$ corresponds to a mapping

$$\mathbf{v} \to \mathbf{w} = A\mathbf{v}$$

of a point of the xy-plane into itself. To find the image of a point P in the plane multiply its coordinate column $\mathbf{v}=[P]$ by A to obtain the coordinate column $\mathbf{w}=[Q]$ of its point of reflection Q.

Example 7.1

Use $A\mathbf{v} = \mathbf{w}$ to find the reflections of the following points in the line $y=x$.

Point P	Matrix Equation $A\mathbf{v}=\mathbf{w}$	Point of Reflection Q
1) $P=(1,0)$	$\begin{pmatrix} 0 & 1 \\ 1 & 0 \end{pmatrix}\begin{pmatrix} 1 \\ 0 \end{pmatrix} = \begin{pmatrix} 0 \\ 1 \end{pmatrix}$	$Q=(0,1)$
2) $P=(0,1)$	$\begin{pmatrix} 0 & 1 \\ 1 & 0 \end{pmatrix}\begin{pmatrix} 0 \\ 1 \end{pmatrix} = \begin{pmatrix} 1 \\ 0 \end{pmatrix}$	$Q=(1,0)$
3) $P=(0,-2)$	$\begin{pmatrix} 0 & 1 \\ 1 & 0 \end{pmatrix}\begin{pmatrix} 0 \\ -2 \end{pmatrix} = \begin{pmatrix} -2 \\ 0 \end{pmatrix}$	$Q=(-2,0)$
4) $P=(2,3)$	$\begin{pmatrix} 0 & 1 \\ 1 & 0 \end{pmatrix}\begin{pmatrix} 2 \\ 3 \end{pmatrix} = \begin{pmatrix} 3 \\ 2 \end{pmatrix}$	$Q=(3,2)$
5) $P=(a,a)$	$\begin{pmatrix} 0 & 1 \\ 1 & 0 \end{pmatrix}\begin{pmatrix} a \\ a \end{pmatrix} = \begin{pmatrix} a \\ a \end{pmatrix}$	$Q=(a,a)$
6) $P=(a,-a)$	$\begin{pmatrix} 0 & 1 \\ 1 & 0 \end{pmatrix}\begin{pmatrix} a \\ -a \end{pmatrix} = \begin{pmatrix} -a \\ a \end{pmatrix}$	$Q=(-a,a)$

In Part 5 of the example, $\mathbf{w} = \mathbf{v}$, hence the reflection of that vector is itself. Geometrically, the vector represents any point on the line $y = x$. In Part 6 $\mathbf{w} = -\mathbf{v}$ which means that the reflection is the negative of the original vector. Geometrically, the vector is perpendicular to the linc. What we have in these two cases are vectors whose reflections are multiples of themselves, i.e. $\mathbf{w} = \lambda\mathbf{v}$ where $\lambda = 1$ or -1. These vectors are solutions of the matrix equation

$$A\mathbf{v} = \lambda\mathbf{v} \quad . \tag{7.1}$$

Note that this equation represents a very special property of the reflection operation. In general $A\mathbf{v} = \mathbf{w}$ where $\mathbf{w} \neq \lambda\mathbf{v}$. It is only for very special vectors $\mathbf{v} \neq 0$ and scalars λ that Equation (7.1) is true.

Let us now interpret Equation (7.1) much more generally. Let A by any $n \times n$ matrix. Then in order to solve Equation (7.1) we need to find a column matrix $\mathbf{v} \neq 0$ and a scalar λ for which the multiplication A times $\mathbf{v}$ equals the scalar λ times $\mathbf{v}$. ***Each λ is called an eigenvalue of the matrix and all non-zero vectors which satisfy Equation (7.1) for that λ are called eigenvectors associated with the eigenvalue λ.***

Equation (7.1) is not in the usual form for a system of equations since the column of unknowns is on both sides of the equation. To obtain the standard form, subtract the term $\lambda\mathbf{v}$ from both sides and rewrite it as $\lambda I\mathbf{v}$ to give it the same matrix structure as the $A\mathbf{v}$ term. Then factoring $\mathbf{v}$ out to the right gives the usual form. Thus we have $A\mathbf{v} = \lambda\mathbf{v} = \lambda I\mathbf{v}$, $A\mathbf{v} = \lambda I\mathbf{v} = 0$, and finally

$$(A - \lambda I)\mathbf{v} = \mathbf{0} \tag{7.2}$$

where $A - \lambda I$ is the coefficient matrix and $\mathbf{0}$ is the zero vector. Therefore, the solution of Equation (7.1) has thus been reduced to the familiar problem of solving an $n \times n$ system of equations with coefficient matrix $A - \lambda I$ and a right hand side zero vector, i.e. a homogeneous system of equations (7.2).

7.2 Calculating Eigenvalues and Eigenvectors

Note that $\mathbf{v} = \mathbf{0}$ is a solution of (7.2) for all λ. It is called the trivial solution and is of no interest, since we are looking for non-zero vector solutions of (7.2). A homogeneous system of equations (such as (7.2)) has non-trivial solutions if and only if the determinant of the coefficient matrix $A - \lambda I$ is zero or $A - \lambda I$ is a singular matrix. We therefore require that

$$\det(A - \lambda I) = 0 \quad . \tag{7.3}$$

This is a polynomial equation and this condition can only be satisfied by at most n values of λ where n is the degree of the polynomial. This equation determines all the eigenvalues of A. The eigenvectors corresponding to a specific eigenvalue λ are then found by explicitly solving Equation (7.2) with that specific λ substituted into the coefficient matrix $A - \lambda I$. The following examples illustrate and discuss these techniques.

Example 7.2 Find the eigenvalues and eigenvectors for the matrix

$$A = \begin{pmatrix} 0 & 1 \\ 1 & 0 \end{pmatrix} .$$

Solution.

First we find the eigenvalues from

$$\det(A - \lambda I) = \begin{vmatrix} -\lambda & 1 \\ 1 & -\lambda \end{vmatrix} = \lambda^2 - 1 = 0 \quad .$$

The eigenvalues are then the roots of the polynomial $(\lambda^2 - 1)$, i.e. 1 and -1.

To find the eigenvectors for the eigenvalue 1, we need to solve the system $(A - \lambda I)\mathbf{v} = \mathbf{0}$ or

$$\begin{pmatrix} -1 & 1 \\ 1 & -1 \end{pmatrix} \begin{pmatrix} x \\ y \end{pmatrix} = \begin{pmatrix} 0 \\ 0 \end{pmatrix} .$$

Row reducing the augmented matrix we get

$$\left(\begin{array}{cc|c} -1 & 1 & 0 \\ 1 & -1 & 0 \end{array} \right) \sim \left(\begin{array}{cc|c} -1 & 1 & 0 \\ 0 & 0 & 0 \end{array} \right) .$$

So we must have $x = y$. Hence the solutions are

$$\mathbf{v} = \begin{pmatrix} a \\ a \end{pmatrix} = a \begin{pmatrix} 1 \\ 1 \end{pmatrix}$$

where a is any non-zero real number.

We have found an infinite collection of eigenvectors associated with the eigenvalue 1, but they are all multiples of one vector. We have found a vector equation for the line through the origin with a direction vector $\langle 1, 1 \rangle$. If we use $x = a = y$ to eliminate $a \neq 0$ we have the points on the line $y = x$ with the origin removed. The eigenvectors corresponding to the eigenvalue -1 turn out to be all non-zero scalar multiples of the vector $\langle 1, -1\rangle^T$ (see *Exercise* 1). These are the position vectors of all the non-zero points on the line $y = -x$. Thus, for each eigenvalue $\lambda = 1$ and $\lambda = -1$, we have obtained vector equations for lines through the origin. The eigenvectors are ***direction vectors*** for these lines. By solving the eigenvalue problem for the matrix representing reflection across the $y = x$ line, we have found that the only vectors reflected into a multiple of themselves are those lying along the axis of reflection and those lying along the line perpendicular to that axis. These results coincide with geometric intuition.

We make the following definitions for an $n \times n$ matrix A.

Definition 7.1 The ***characteristic polynomial*** for A, $f_A(\lambda)$, is $(-1)^n \det(A - \lambda I)$ which in expanded form is $\lambda^n - c_1\lambda^{n-1} + c_2\lambda^{n-2} - \cdots + (-1)^n c_n$.

Definition 7.2 The ***characteristic equation*** for A is $\det(A - \lambda I) = 0$ or $f_A(\lambda) = 0$.

Definition 7.3 The ***eigenvalues*** of A are the roots λ_i of $f_A(\lambda)$ the characteristic polynomial of A. They are the solutions of the characteristic equation for A.

Definition 7.4 Any **non-zero** vector $\mathbf{v}$ which satisfies the vector equation $(A - \lambda I)\mathbf{v} = \mathbf{0}$, i.e. $A\mathbf{v} = \lambda\mathbf{v}$ is called an ***eigenvector*** corresponding to the eigenvalue λ.

Note: If $\mathbf{v}$ is an eigenvector corresponding to the eigenvalue λ, then $c\mathbf{v}$, where c is any non-zero real number, is also an eigenvector corresponding to the same eigenvalue (see *Exercise* 14). Geometrically, this corresponds to the fact that an eigenvector is a direction vector for a straight line through the origin. Any non-zero vector on a line through the origin can serve as a direction vector for that line.

For a 2×2 matrix

$$A = \begin{pmatrix} a_{11} & a_{12} \\ a_{21} & a_{22} \end{pmatrix}$$

its characteristic equation is given by

$$f_A(\lambda) = \lambda^2 - c_1\lambda + c_2 = \det(A - \lambda I) = \begin{vmatrix} a_{11} - \lambda & a_{12} \\ a_{21} & a_{22} - \lambda \end{vmatrix}$$

$$= \lambda^2 - (a_{11} + a_{22})\lambda + a_{11}a_{22} - a_{12}a_{21} = 0 \quad .$$

Here the coefficient $c_1 = a_{11} + a_{22}$ is the sum of the diagonal terms of A and the coefficient $c_2 = \det A$ equals the determinant of A.

For a 3×3 matrix

$$A = \begin{pmatrix} a_{11} & a_{12} & a_{13} \\ a_{21} & a_{22} & a_{23} \\ a_{31} & a_{32} & a_{33} \end{pmatrix}$$

its characteristic equation is given by

$$f_A(\lambda) = \lambda^3 - c_1\lambda^2 + c_2\lambda - c_3 = (-1)^3 \det(A - \lambda I) = -\det(A - \lambda I) = 0$$

where

$$\begin{aligned} c_1 &= a_{11} + a_{22} + a_{33} \quad , \\ c_2 &= \begin{vmatrix} a_{11} & a_{12} \\ a_{21} & a_{22} \end{vmatrix} + \begin{vmatrix} a_{11} & a_{13} \\ a_{31} & a_{33} \end{vmatrix} + \begin{vmatrix} a_{22} & a_{23} \\ a_{32} & a_{33} \end{vmatrix} \\ &= (a_{11}a_{22} - a_{12}a_{21}) + (a_{11}a_{33} - a_{13}a_{31}) + (a_{22}a_{33} - a_{23}a_{32}) \quad , \end{aligned}$$

and

$$c_3 = \det A \quad .$$

In general if A is an $n \times n$ matrix the ***trace of*** A (denoted by TrA) is defined to be the sum $\text{Tr}A = a_{11} + a_{22} + \cdots + a_{nn}$ of its diagonal terms. If $f_A(\lambda) = \lambda^n - c_1\lambda^{n-1} + \cdots + (-1)^n c_n$ is the characteristic polynomial of A then $c_1 = \text{Tr}A = a_{11} + a_{22} + \cdots + a_{nn}$ and $c_n = \det A$. For $k = 1, 2, \ldots, n$ the terms c_k equal the sums of the ***principal*** $k \times k$ ***minors of*** A . They were given above for the cases $n = 2$ and $n = 3$. In general a $k \times k$ minor of A is one of the $k \times k$ determinants

$$\begin{vmatrix} a_{i_1 i_1} & \cdots & a_{i_1 i_k} \\ \vdots & & \vdots \\ a_{i_k i_1} & \cdots & a_{i_k i_k} \end{vmatrix}$$

where $1 \leq i_1 < \cdots < i_k \leq n$.

If $\lambda_1, \lambda_2, \ldots \lambda_n$ are the n eigenvalues of A (repetitions are possible) then $f_A(\lambda) = (\lambda - \lambda_1)(\lambda - \lambda_2)\cdots(\lambda - \lambda_n) = \lambda^n - c_1\lambda^{n-1} + \cdots + (-1)^n c_n$ and it can be shown that $\text{Tr}A = c_1 = \lambda_1 + \lambda_2 + \cdots + \lambda_n$ and that $\det A = c_n = \lambda_1\lambda_2\ldots\lambda_n$.

Example 7.3 Find the eigenvalues and associated eigenvectors for A if

$$A = \begin{pmatrix} -3 & 1 & 0 \\ 1 & -2 & 1 \\ 0 & 1 & -3 \end{pmatrix} \quad .$$

Solution.

We will first find the characteristic polynomial for A by directly computing $(-1)^3 \det(A - \lambda I)$.

$$f_A(\lambda) = (-1)^3 \begin{vmatrix} -3-\lambda & 1 & 0 \\ 1 & -2-\lambda & 1 \\ 0 & 1 & -3-\lambda \end{vmatrix}$$

$$= (\lambda+3) \begin{vmatrix} -2-\lambda & 1 \\ 1 & -3-\lambda \end{vmatrix} + \begin{vmatrix} 1 & 1 \\ 0 & -3-\lambda \end{vmatrix}$$

$$= (\lambda+3)(\lambda^2+5\lambda+5) + (-3-\lambda) = (\lambda+3)(\lambda^2+5\lambda+4) =$$

$$(\lambda+3)(\lambda+1)(\lambda+4) = \lambda^3 + 8\lambda^2 + 19\lambda + 12 \quad .$$

So the characteristic equation is

$$\lambda^3 + 8\lambda^2 + 19\lambda + 12 = (\lambda+3)(\lambda+1)(\lambda+4) = 0 \quad .$$

and so the eigenvalues of A are $\lambda_1 = -1$, $\lambda_2 = -3$, and $\lambda_3 = -4$.

If we use the sums of the principal $k \times k$ minors to find c_k for $k = 1, 2, 3$ then

$$\text{Tr}A = c_1 = -8$$

$$c_2 = \begin{vmatrix} -3 & 1 \\ 1 & -2 \end{vmatrix} + \begin{vmatrix} -3 & 0 \\ 0 & -3 \end{vmatrix} + \begin{vmatrix} -2 & 1 \\ 1 & -3 \end{vmatrix} = 5 + 9 + 5 = 19$$

and

$$\det A = c_3 = \begin{vmatrix} -3 & 1 & 0 \\ 1 & -2 & 1 \\ 0 & 1 & -3 \end{vmatrix} = \begin{vmatrix} 0 & -5 & 3 \\ 1 & -2 & 1 \\ 0 & 1 & -3 \end{vmatrix} = -\begin{vmatrix} -5 & 3 \\ 1 & -3 \end{vmatrix} = -12 \quad .$$

Thus, $f_A(\lambda) = \lambda^3 - c_1\lambda^2 + c_2\lambda - c_3 = \lambda^3 + 8\lambda^2 + 19\lambda + 12$ in agreement with our previous calculation.

Observe that $c_1 = -8 = -1 - 3 - 4 = \lambda_1 + \lambda_2 + \lambda_3$ and that $c_3 = -12 = (-1)(-3)(-4) = \lambda_1\lambda_2\lambda_3$.

We will now find the eigenvectors corresponding to $\lambda_1 = -1$. Analysis of the system of equations $(A - \lambda I)\mathbf{v} = \mathbf{0}$ or

$$\begin{pmatrix} -2 & 1 & 0 \\ 1 & -1 & 1 \\ 0 & 1 & -2 \end{pmatrix} \begin{pmatrix} x \\ y \\ z \end{pmatrix} = \begin{pmatrix} 0 \\ 0 \\ 0 \end{pmatrix}$$

yields $x = z$ and $y = 2z$. Hence the eigenvectors are

$$\mathbf{v} = \begin{pmatrix} a \\ 2a \\ a \end{pmatrix} = a \begin{pmatrix} 1 \\ 2 \\ 1 \end{pmatrix}$$

where a is an arbitrary non-zero real number. By working Exercise 2 in a similar manner, you will find the eigenvectors for the other eigenvalues to be given by

$$b \begin{pmatrix} 1 \\ 0 \\ -1 \end{pmatrix} \quad \text{for} \quad \lambda_2 = -3 \quad \text{and} \quad c \begin{pmatrix} 1 \\ -1 \\ 1 \end{pmatrix} \quad \text{for} \quad \lambda_3 = -4$$

where $b \neq 0 \neq c$.

In Examples 7.2 and 7.3 the eigenvalues were distinct (*all different*), but this does not happen for every matrix as illustrated in the next example.

Example 7.4 Calculate the eigenvalues and eigenvectors for

$$A = \begin{pmatrix} 0 & 2 & 2 \\ 2 & 0 & 2 \\ 2 & 2 & 0 \end{pmatrix} .$$

Solution.

Using the $k \times k$ principal minors to find c_1, c_2, and c_3 we have

$$c_1 = 0 + 0 + 0 = 0$$

$$c_2 = \begin{vmatrix} 0 & 2 \\ 2 & 0 \end{vmatrix} + \begin{vmatrix} 0 & 2 \\ 2 & 0 \end{vmatrix} + \begin{vmatrix} 0 & 2 \\ 2 & 0 \end{vmatrix} = -4 - 4 - 4 = -12$$

and

$$c_3 = \begin{vmatrix} 0 & 2 & 2 \\ 2 & 0 & 2 \\ 2 & 2 & 0 \end{vmatrix} = \begin{vmatrix} 0 & 2 & 2 \\ 0 & -2 & 2 \\ 2 & 2 & 0 \end{vmatrix} = 2 \begin{vmatrix} 2 & 2 \\ -2 & 2 \end{vmatrix} = 2(8) = 16 \quad .$$

Thus, $f_A(\lambda) = \lambda^3 - c_1\lambda^2 + c_2\lambda - c_3 = \lambda^3 - 12\lambda - 16$ so the characteristic equation is

$$(-1)^3 \det(A - \lambda I) = (\lambda + 2)^2(\lambda - 4) = 0 \quad .$$

Hence $\lambda_1 = -2$, $\lambda_2 = -2$, and $\lambda_3 = 4$ and A has a repeated eigenvalue -2. For this repeated eigenvalue the corresponding eigenvectors are solutions of the equation $(A + 2I)\mathbf{v} = \mathbf{0}$. Row reduction of the augmented matrix gives

$$\left(\begin{array}{ccc|c} 2 & 2 & 2 & 0 \\ 2 & 2 & 2 & 0 \\ 2 & 2 & 2 & 0 \end{array}\right) \sim \left(\begin{array}{ccc|c} 2 & 2 & 2 & 0 \\ 0 & 0 & 0 & 0 \\ 0 & 0 & 0 & 0 \end{array}\right) \sim \left(\begin{array}{ccc|c} 1 & 1 & 1 & 0 \\ 0 & 0 & 0 & 0 \\ 0 & 0 & 0 & 0 \end{array}\right) \quad .$$

Thus, all eigenvectors $(x, y, z)^T$ corresponding to $\lambda = -2$ satisfy $x + y + z = 0$. So $x = -y - z$ is required of them, i.e.

$$\mathbf{v} = \begin{pmatrix} x \\ y \\ z \end{pmatrix} = \begin{pmatrix} -a - b \\ a \\ b \end{pmatrix} = \begin{pmatrix} -a \\ a \\ 0 \end{pmatrix} + \begin{pmatrix} -b \\ 0 \\ b \end{pmatrix} = a \begin{pmatrix} -1 \\ 1 \\ 0 \end{pmatrix} + b \begin{pmatrix} -1 \\ 0 \\ 1 \end{pmatrix} \quad .$$

Thus the eigenvectors corresponding to the eigenvalue $\lambda = -2$ of multiplicity two are linear combinations (*sums of scalar multiples*) of the two vectors

$$\mathbf{v}_1 = \begin{pmatrix} -1 \\ 1 \\ 0 \end{pmatrix} \quad \text{and} \quad \mathbf{v}_2 = \begin{pmatrix} -1 \\ 0 \\ 1 \end{pmatrix} \quad .$$

Only the choice $a = b = 0$ is excluded to avoid the trivial solution. The condtion $x + y + z = 0$ is the equation of a plane, so geometrically, the collection of eigenvectors corresponding to the eigenvalue $\lambda = -2$ belong to this plane through the origin (*the origin is excluded*). A specific eigenvector is obtained by simply choosing values for a and b. Some examples are

$$\mathbf{v}_1 + 0\mathbf{v}_2 = \begin{pmatrix} -1 \\ 1 \\ 0 \end{pmatrix}, \quad 4\mathbf{v}_1 - 3\mathbf{v}_2 = \begin{pmatrix} -1 \\ 4 \\ -3 \end{pmatrix}, \quad \mathbf{v}_1 - \mathbf{v}_2 = \begin{pmatrix} 0 \\ 1 \\ -1 \end{pmatrix} \quad .$$

The eigenvectors corresponding to $\lambda_3 = 4$ turn out to be non-zero scalar multiples of $\mathbf{v}_3 = \begin{pmatrix} 1 \\ 1 \\ 1 \end{pmatrix}$ and correspond to the line through the origin in $\mathfrak{R}^3$ which is perpendicular to the plane $x + y + z = 0$.

7.3 The Existence of a Complete Set of Mutually Orthogonal Eigenvectors (The Symmetric Matrix Case)

By calculating the dot product of any two eigenvectors belonging to two distinct eigenvalues in Example 7.3, you will find that its value is zero. Thus two such eigenvectors are orthogonal (perpendicular) to each other. (see *Exercise* 15.) Likewise, in Example 7.4, we find $\mathbf{v}_1 \bullet \mathbf{v}_3 = 0$ and $\mathbf{v}_2 \bullet \mathbf{v}_3 = 0$ so that eigenvectors corresponding to distinct eigenvalues are again found to be orthogonal. (See *Exercise* 16).

At this point we might guess that for any matrix, eigenvectors belonging to distinct eigenvalues are orthogonal. The following example shows that this is unfortunately not the case in general.

Example 7.5 The eigenvalues and eigenvectors for the matrix

$$A = \begin{pmatrix} 1 & 1 \\ -2 & 4 \end{pmatrix}$$

are found in the usual way. The eigenvalues are $\lambda_1 = 2$ and $\lambda_2 = 3$. The eigenvector

$$\mathbf{v}_1 = \begin{pmatrix} 1 \\ 1 \end{pmatrix}$$

corresponds to $\lambda_1 = 2$ and the eigenvector

$$\mathbf{v}_2 = \begin{pmatrix} 1 \\ 2 \end{pmatrix}$$

corresponds to $\lambda_2 = 3$.

But $\mathbf{v}_1 \bullet \mathbf{v}_2 = 3$, so the eigenvectors are not orthogonal even though they belong to distinct eigenvalues. This cannot happen for a symmetric matrix.

Definition 7.5 An $n \times n$ matrix A is symmetric if $A^T = A$ or $a_{ji} = a_{ij}$ for all $i, j = 1, \cdots, n$.

According to the following theorem, for symmetric matrices eigenvectors corresponding to distinct eigenvalues must be orthogonal.

Theorem 7.1. *If A is a symmetric $n \times n$ matrix, $\lambda_1 \neq \lambda_2$ are two distinct eigenvalues, and if $\mathbf{v}_1$ and $\mathbf{v}_2$ are corresponding eigenvectors then $\mathbf{v}_1 \bullet \mathbf{v}_2 = 0$. (Symbolically if $A = A^T$, $A\mathbf{v}_1 = \lambda_1\mathbf{v}_1$ and $A\mathbf{v}_2 = \lambda_2\mathbf{v}_2$ where $\mathbf{v}_1 \neq \mathbf{0} \neq \mathbf{v}_2$ and $\lambda_1 \neq \lambda_2$ then $\mathbf{v}_1 \bullet \mathbf{v}_2 = 0$.)*

This theorem guarantees that for a symmetric matrix, eigenvectors corresponding to distinct eigenvalues must be orthogonal. The matrices in Examples 7.2, 7.3, and 7.4 illustrate this property. Symmetric matrices are part of a larger class of matrices (*called* ***normal*** *matrices*) which have orthogonal eigenvectors corresponding to distinct eigenvalues.

The eigenvalue problem arises in the principal axis problem discussed in Chapter 8. There we will need to find n orthogonal eigenvectors for an $n \times n$ symmetric matrix A. Theorem 7.1 guarantees their existence if A happens to have n distinct eigenvalues, but in general A could have repeated eigenvalues. Theorem 7.1 can not be used to guarantee the automatic existence of n mutually orthogonal eigenvectors when some of the eigenvalues are repeated. However, the following theorem gives us the desired guarantee.

Theorem 7.2. *Every $n \times n$ symmetric matrix has n mutually orthogonal eigenvectors.*

This situation is also expressed by saying that A has a ***complete set*** of mutually orthogonal eigenvectors. Theorem 7.2 states that the $n \times n$ symmetric matrix A has a complete set but does not say how to find them. (*In general the $n \times n$ matrices with n mutually orthogonal eigenvectors are called* ***normal matrices***.)

Example 7.4 is a case where a 3×3 symmetric matrix A does not have all distinct eigenvalues. The repeated eigenvalue has a corresponding collection of eigenvectors with two degrees of freedom. They lie in a plane and two linearly independent eigenvectors are chosen. The two eigenvectors we select are not perpendicular to each other but they are perpendicular to the eigenvector corresponding to the other eigenvalue. Specifically,

$$\mathbf{v}_1 \bullet \mathbf{v}_2 = \langle -1, 1, 0 \rangle \bullet \langle -1, 0, 1 \rangle = 1$$

How can we find the promised three mutually orthogonal eigenvectors?

We can appeal to geometry for the answer to this question. Let us consider the three eigenvectors $\mathbf{v}_1$, $\mathbf{v}_2$, and $\mathbf{v}_3$ from Example 7.4 as position vectors in $\mathcal{R}^3$. Recall that the set of eigenvectors corresponding to the eigenvalue -2 is the set of all non-zero linear combinations of $\mathbf{v}_1$ and $\mathbf{v}_2$ and forms a plane containing the origin. Since $\mathbf{v}_3$ is perpendicular to both $\mathbf{v}_1$ and $\mathbf{v}_2$, it is also perpendicular to the plane they form. In fact a vector $\mathbf{v}$ is in this plane if and only if $\mathbf{v}$ is orthogonal to $\mathbf{v}_3$. Now consider the vector $\mathbf{v'}_1 = \mathbf{v}_2 \times \mathbf{v}_3$ as a replacement for $\mathbf{v}_1$. Since the cross product of two vectors is a vector perpendicular to each of the two vectors involved, we have $\mathbf{v'}_1$ perpendicular to both $\mathbf{v}_2$ and $\mathbf{v}_3$, so these three form a mutually orthogonal set of vectors. But the non-zero vector $\mathbf{v'}_1 = \mathbf{v}_2 \times \mathbf{v}_3$ is orthogonal to $\mathbf{v}_3$ and must lie in the plane formed by $\mathbf{v}_1$ and $\mathbf{v}_2$ (*see Figure 7.2*). Thus it is an eigenvector belonging to the eigenvalue -2.

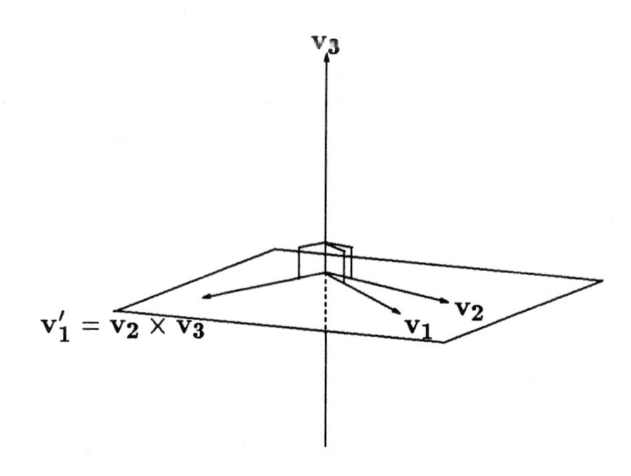

Figure 7.2

Thus $\mathbf{v'}_1$, $\mathbf{v}_2$, and $\mathbf{v}_3$ form a complete set of mutually orthogonal eigenvectors for the symmetric matrix A in Example 7.4. Specifically,

$$\mathbf{v'}_1 = \begin{pmatrix} -1 \\ 2 \\ -1 \end{pmatrix} \qquad \mathbf{v}_2 = \begin{pmatrix} -1 \\ 0 \\ 1 \end{pmatrix} \qquad \mathbf{v}_3 = \begin{pmatrix} 1 \\ 1 \\ 1 \end{pmatrix} .$$

7.4 Exercises

1. Show that any eigenvector corresponding to $\lambda = -1$ in Example 7.2 is a scalar multiple of the vector $\langle 1, -1 \rangle^T$.

2. Show that any eigenvector corresponding to $\lambda_3 = -4$ in Example 7.3 is a scalar multiple of the vector $\langle 1, -1, 1 \rangle^T$.

3. Find the characteristic polynomials

$$f_A(\lambda) = \lambda^n - c_1\lambda^{n-1} + \cdots + (-1)^n c_n = (-1)^n \det(A - \lambda I)$$

for each of the following matrices:

(a) $\begin{pmatrix} 3 & 1 \\ 2 & -1 \end{pmatrix}$ **(b)** $\begin{pmatrix} 2 & -1 \\ 1 & 3 \end{pmatrix}$ **(c)** $\begin{pmatrix} -2 & 1 \\ 5 & -3 \end{pmatrix}$ **(d)** $\begin{pmatrix} -2 & 3 \\ 5 & -1 \end{pmatrix}$

(e) $\begin{pmatrix} 2 & 1 \\ 5 & -2 \end{pmatrix}$ **(f)** $\begin{pmatrix} 2 & -1 \\ 13 & -2 \end{pmatrix}$ **(g)** $\begin{pmatrix} 1 & 2 & 3 \\ -1 & 1 & 2 \\ 2 & -1 & 3 \end{pmatrix}$

(h) $\begin{pmatrix} 2 & -1 & 2 \\ 3 & 1 & 1 \\ -1 & 2 & -3 \end{pmatrix}$ **(i)** $\begin{pmatrix} 1 & 2 & -1 \\ 1 & -2 & 2 \\ 1 & 1 & -1 \end{pmatrix}$ **(j)** $\begin{pmatrix} 1 & -1 & 1 \\ 2 & 1 & 3 \\ 3 & -2 & 4 \end{pmatrix}$.

4. Find the eigenvalues for each of the matrices:

a) $\begin{pmatrix} 4 & 2 \\ 2 & 1 \end{pmatrix}$ **b)** $\begin{pmatrix} 5 & -2 \\ -2 & 2 \end{pmatrix}$ **c)** $\begin{pmatrix} 3 & 1 & 1 \\ 1 & 0 & 2 \\ 1 & 2 & 0 \end{pmatrix}$ **d)** $\begin{pmatrix} 3 & 1 & 4 \\ 0 & 2 & 6 \\ 0 & 0 & 5 \end{pmatrix}$

e) $\begin{pmatrix} 1 & 0 \\ 2 & 1 \end{pmatrix}$ **f)** $\begin{pmatrix} 3 & -1 \\ 1 & 1 \end{pmatrix}$ **g)** $\begin{pmatrix} 5 & 0 & 0 \\ 0 & 4 & 2 \\ 0 & 2 & 1 \end{pmatrix}$ **h)** $\begin{pmatrix} 2 & 0 & 0 \\ 0 & 1 & -1 \\ 0 & 2 & 4 \end{pmatrix}$.

5. Find an eigenvector corresponding to each eigenvalue for the matrices in Exercise 4.

6. If A is the matrix

$$A = \begin{pmatrix} -3 & 4 & 3 \\ 4 & 5 & 1 \\ 3 & 1 & 12 \end{pmatrix}$$

and if

$$\mathbf{u} = \begin{pmatrix} 1 \\ -1 \\ 3 \end{pmatrix} \quad \text{and} \quad \mathbf{v} = \begin{pmatrix} 1 \\ 3 \\ -1 \end{pmatrix}$$

compute $A\mathbf{u}$ and $A\mathbf{v}$ to determine whether or not $\mathbf{u}$ and $\mathbf{v}$ are eigenvectors of A.

7. Repeat Problem 6 for the following matrices

a) $A = \begin{pmatrix} 3 & -1 & 3 \\ -1 & 7 & -1 \\ 3 & -1 & 3 \end{pmatrix}$, $\mathbf{u} = \begin{pmatrix} 1 \\ -1 \\ -1 \end{pmatrix}$, $\mathbf{v} = \begin{pmatrix} 1 \\ -2 \\ 1 \end{pmatrix}$

b) $A = \begin{pmatrix} 5 & -2 & 2 \\ -2 & 2 & -1 \\ 2 & -1 & 2 \end{pmatrix}$, $\mathbf{u} = \begin{pmatrix} 1 \\ -1 \\ 2 \end{pmatrix}$, $\mathbf{v} = \begin{pmatrix} 2 \\ -1 \\ 1 \end{pmatrix}$.

8. Use direct multiplication to show that, for each of the following matrices A, the given vectors $\mathbf{v}_1$, $\mathbf{v}_2$, and $\mathbf{v}_3$ are eigenvectors of A and to find the eigenvalues λ_1, λ_2, and λ_3 of A.

a) $A = \begin{pmatrix} 2 & 1 & 3 \\ -1 & 6 & -1 \\ 3 & -5 & 2 \end{pmatrix}, \mathbf{v}_1 = \begin{pmatrix} 1 \\ 0 \\ -1 \end{pmatrix}, \mathbf{v}_2 = \begin{pmatrix} 2 \\ 1 \\ 1 \end{pmatrix}, \mathbf{v}_3 = \begin{pmatrix} 7 \\ -9 \\ 11 \end{pmatrix}$

b) $A = \begin{pmatrix} 6 & -1 & 3 \\ -1 & 10 & -1 \\ 3 & -1 & 6 \end{pmatrix}, \mathbf{v}_1 = \begin{pmatrix} 1 \\ -2 \\ 1 \end{pmatrix}, \mathbf{v}_2 = \begin{pmatrix} 1 \\ 1 \\ 1 \end{pmatrix}, \mathbf{v}_3 = \begin{pmatrix} -1 \\ 0 \\ 1 \end{pmatrix}.$

9. The eigenvalues of

$$A = \begin{pmatrix} 3 & 3 & 2 \\ 3 & 6 & 5 \\ 2 & 5 & 11 \end{pmatrix}$$

are $\lambda_1 = 15,\ \lambda_2 = 4$, and $\lambda_3 = 1$.

a) Find a unit eigenvector $\mathbf{q}_1$ corresponding to λ_1.

b) Find a unit eigenvector $\mathbf{q}_2$ corresponding to λ_2.

c) Find a unit eigenvector $\mathbf{q}_3$ corresponding to λ_3. and such that $\{\mathbf{q}_1, \mathbf{q}_2, \mathbf{q}_3\}$ is a right handed set of vectors.

10. Repeat Problem 9 for the matrix

$$A = \begin{pmatrix} 11 & -3 & -2 \\ -3 & 8 & -5 \\ -2 & -5 & 3 \end{pmatrix}$$

which has eigenvalues $\lambda_1 = -1,\ \lambda_2 = 10$, and $\lambda_3 = 13$.

11. Find the eigenvalues and corresponding eigenvectors of the following matrices

a) $\begin{pmatrix} 2 & -3 & 2 \\ 1 & 0 & 0 \\ 2 & -1 & 0 \end{pmatrix}$ **b)** $\begin{pmatrix} 1 & 1 & 1 \\ 1 & -1 & -1 \\ 1 & -1 & 1 \end{pmatrix}.$

12. Find the eigenvalues and corresponding ***mutually orthogonal*** eigenvectors of the following matrices

a) $\begin{pmatrix} 0 & -4 & 4 \\ -4 & -6 & 8 \\ 4 & 8 & -6 \end{pmatrix}$ **b)** $\begin{pmatrix} 4 & 1 & 1 \\ 1 & 4 & 1 \\ 1 & 1 & 4 \end{pmatrix}.$

13. Show that the eigenvectors are mutually orthogonal for the following matrices from Exercise 4 and Exercise 5

a) **b)** **c)** **g)** .

14. Prove that if $\mathbf{v}$ is an eigenvector corresponding to an eigenvalue λ of the matrix A, then for $c \neq 0$ the vector $c\mathbf{v}$ is also an eigenvector corresponding to λ.

15. For Example 7.3 verify that ***any*** two eigenvectors corresponding to distinct eigenvalues are orthogonal.

16. For Example 7.4 verify that ***any*** two eigenvectors corresponding to distinct eigenvalues are orthogonal.

17. For Example 7.4 show that if we let $\mathbf{v}' = \mathbf{v}_2 \times \mathbf{v}_3$ and $\mathbf{x}$ is any eigenvector expressible as $\mathbf{x} = a\mathbf{v}_1 + b\mathbf{v}_2$, then we can find constants r and s such that $\mathbf{x} = r\mathbf{v}_2 + s\mathbf{v}'$.

18. For the matrix of Exercise 4g find three mutually orthogonal eigenvectors.

19. The matrix $\begin{pmatrix} 7 & -2 & 1 \\ -2 & 10 & -2 \\ 1 & -2 & 7 \end{pmatrix}$ has eigenvalues 6, 6, and 12. Find three corresponding mutually orthogonal eigenvectors.

20. Repeat Problem 19 for the matrix

$$A = \begin{pmatrix} 7 & -2 & 4 \\ -2 & 4 & -2 \\ 4 & -2 & 7 \end{pmatrix}$$

which has the eigenvalues 3,3, and 12.

21. The numbers $\lambda_3 = \lambda_1 = 20$ are repeated eigenvalues of the matrix

$$A = \begin{pmatrix} 11 & 6 & -15 \\ 6 & 16 & 10 \\ -15 & 10 & -5 \end{pmatrix} .$$

a) Find a unit eigenvector $\mathbf{q}_2$ corresponding to $\lambda_2 \neq 20$ without finding λ_2.

b) Then use matrix multiplication to find λ_2 from $A\mathbf{q}_2 = \lambda_2\mathbf{q}_2$ and to verify that your $\mathbf{q}_2$ from **(a)** is correct.

c) The vector $\mathbf{q}_3 = \frac{1}{\sqrt{3}}[1, -1, 1]^T$ is one unit eigenvector of A corresponding to $\lambda_3 = 20$. Determine another unit eigenvector $\mathbf{q}_1$ corresponding to $\lambda_1 = 20$ such that $\{\mathbf{q}_1, \mathbf{q}_2, \mathbf{q}_3\}$ is a right handed set of orthonormal vectors. Directly verify by matrix multiplication that $A\mathbf{q}_1 = 20\mathbf{q}_1$.

22. If A is the matrix

$$A = \begin{pmatrix} 3 & -1 & 2 \\ -1 & 3 & -2 \\ 2 & -2 & 6 \end{pmatrix}$$

then $\lambda_2 = \lambda_3 = 2$ is a twice repeated eigenvalue of A.

a) Find a unit eigenvector $\mathbf{q}_1$ corresponding to $\lambda_1 \neq 2$ without finding λ_1.

b) Use matrix multiplication to find λ_1 from $A\mathbf{q}_1 = \lambda_1\mathbf{q}_1$ and to verify that your $\mathbf{q}_1$ from **(a)** is correct.

c) The vector $\mathbf{q}_3 = \frac{1}{\sqrt{3}}[-1, 1, 1]^T$ is one unit eigenvector corresponding to $\lambda_3 = 2$. Determine another unit eigenvector $\mathbf{q}_2$ corresponding to $\lambda_2 = 2$ such that $\{\mathbf{q}_1, \mathbf{q}_2, \mathbf{q}_3\}$ is a right handed set of orthonormal vectors.

23. Show that if A is a 3×3 matrix, and $\mathbf{v}_1$, $\mathbf{v}_2$, and $\mathbf{v}_3$ are 3×1 column vectors then $A[\mathbf{v}_1, \mathbf{v}_2, \mathbf{v}_3] = [A\mathbf{v}_1, A\mathbf{v}_2, A\mathbf{v}_3]$ where $[\mathbf{v}_1, \mathbf{v}_2, \mathbf{v}_3]$ is the 3×3 matrix with the indicated columns.

24. By using $\det(A - \lambda I)$, show that if A is upper or lower triangular, its main diagonal elements are its eigenvalues.

25. By using $A\mathbf{v} = \lambda\mathbf{v}$, **(a)** show that if A has an inverse and $\lambda \neq 0$ is an eigenvalue of A, then λ^{-1} is an eigenvalue of A^{-1}. **(b)** show that $\alpha\lambda + \beta$ is an eigenvalue of the matrix $\alpha A + \beta I$.

26. By comparing the characteristic polynomials of A and A^T, show that they have the same eigenvalues.

27. Assuming that the characteristic polynomial of a matrix A can be factored, choose a value for λ which will allow you to conclude that the determinant of A equals the product of its eigenvalues.

8 The Principal Axes Problem

Outline of Key Ideas

The restricted second degree equation in two variables:

quadratic form

conic section curves

solution involves orthogonal matrix rotation and eigenvalue problem

summary of the method.

The general second degree equation in two variables:

linear terms

translation of axes.

The general second degree equation in three variables.

The Principal Axes Problem

The equation for a conic section curve in the plane or a quadric surface in three dimensions is simplest to identify and graph when the coordinate axes are chosen to coincide with the natural axes of the curve or surface, their ***principal axes***. If the variables used in the equation for the curve or surface do not correspond to the principal axes, it is desirable to rotate the coordinate axes to change to variables that do.

We will find that the eigenvalue problem of Chapter 7 is needed to discover the principal axes, and the rotation matrix results of Chapter 6 are needed to rotate the present axes into the principal axes.

8.1 The Restricted Second Degree Equation in Two Variables

We will first consider the problem in two variables with the equation containing the variables only in terms of second degree. The more general case when the variables can also occur in first degree terms in the equation will be considered later.

Such an equation has the form

$$ax^2 + 2bxy + cy^2 + f = 0 \quad , \tag{8.1}$$

where x and y are the variables and a, b, c, and f are constants. It is a fact that each equation of this form represents a conic section curve centered at the origin, and that the x and y axes are not the principal axes of the curve if b is non-zero. The sum of the second degree terms is called a ***quadratic form in** x **and** y*.

Finding the graph of the conic section curve represented by a specific equation of this type is fairly easy if $b = 0$ because the equation can be arranged (with minor algebra) into one of the following standard forms:

$$x^2 + y^2 = r^2 \quad \text{a circle with radius } r \tag{8.2}$$

$$\frac{x^2}{p^2} + \frac{y^2}{q^2} = 1 \quad \text{an ellipse with semi-axes lengths } p \text{ and } q \tag{8.3}$$

$$\frac{x^2}{p^2} - \frac{y^2}{q^2} = 1 \quad \text{hyperbola with asymptotes } y = \pm\frac{q}{p}x \quad . \tag{8.4}$$

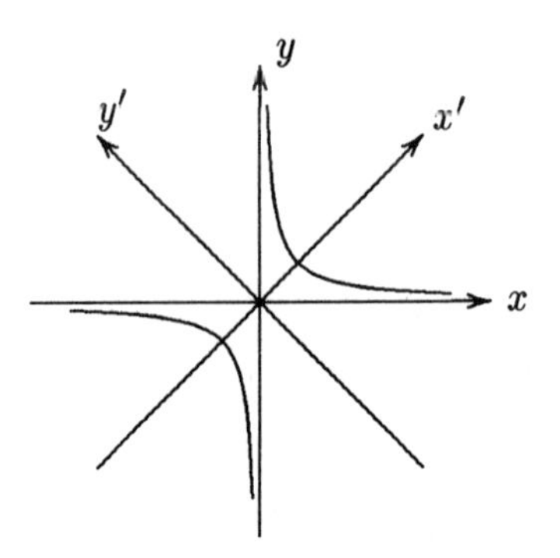

Figure 8.1

Since the x, y axes are the principal axes of the curve in these cases, the specific information on type of curve and dimensions available from these equations make sketching the graph easy.

To illustrate the nature of our problem, consider the very familiar function $y = 1/x$, the graph of which is shown in Figure 8.1. When written as

$$xy - 1 = 0 \quad . \tag{8.5}$$

It has the form of Equation (8.1) with $a = c = 0$, $b = \frac{1}{2}$, and $f = -1$.

So it is a conic section curve and it turns out to be a hyperbola. Notice that its equation is not of the standard form (8.4) since the x and y axes are not the principal axes of the hyperbola. They are in fact the lines $y = x$ and $y = -x$. If the coordinate axes are rotated onto these lines, then the equation for the hyperbola becomes

$$x'^2 - y'^2 = 2 \quad . \tag{8.6}$$

It is the task of this chapter to develop methods for obtaining new x' and y' axes and the resulting equation for a curve from its original equation.

The key geometrical idea is to use a rotation to go from the x, y axes to the x', y' axes. From Chapter 6 we learned that an orthogonal matrix Q with $\det Q = 1$, a rotation matrix, is used so that

$$\mathbf{v} = Q\mathbf{v}' \tag{8.7}$$

where

$$\mathbf{v}' = \begin{pmatrix} x' \\ y' \end{pmatrix} \quad \text{and} \quad \mathbf{v} = \begin{pmatrix} x \\ y \end{pmatrix} \quad . \tag{8.8}$$

Now equation (8.1) can be written in terms of matrices as

$$(8.9) \qquad -f = ax^2 + 2bxy + cy^2 = \mathbf{v}^T A\mathbf{v} = \begin{pmatrix} x & y \end{pmatrix} \begin{pmatrix} a & b \\ b & c \end{pmatrix} \begin{pmatrix} x \\ y \end{pmatrix}$$

where $A = \begin{pmatrix} a & b \\ b & c \end{pmatrix}$ is the symmetric matrix corresponding to this equation. To show that $-f = \mathbf{v}^T A\mathbf{v}$ write (8.9) as follows

$$(8.10) \qquad \begin{aligned} -f &= (ax+by)x + (bx+cy)y = \begin{pmatrix} ax+by & bx+cy \end{pmatrix} \begin{pmatrix} x \\ y \end{pmatrix} \\ &= \begin{pmatrix} x & y \end{pmatrix} \begin{pmatrix} a & b \\ b & c \end{pmatrix} \begin{pmatrix} x \\ y \end{pmatrix} . \end{aligned}$$

For example, Equation (8.5) becomes

$$(8.11) \qquad 1 = xy = \mathbf{v}^T A\mathbf{v} = \begin{pmatrix} x & y \end{pmatrix} \begin{pmatrix} 0 & \frac{1}{2} \\ \frac{1}{2} & 0 \end{pmatrix} \begin{pmatrix} x \\ y \end{pmatrix} .$$

The matrix A contains the coefficients of the quadratic form. In this example

$$(8.12) \qquad A = \begin{pmatrix} 0 & \frac{1}{2} \\ \frac{1}{2} & 0 \end{pmatrix} .$$

Substituting this into (8.9) we obtain

$$-f = (Q\mathbf{v}')^T A(Q\mathbf{v}') = \mathbf{v}'^T Q^T AQ\mathbf{v}' = \mathbf{v}'^T (Q^T AQ)\mathbf{v}'$$

$$(8.13) \qquad -f = \begin{pmatrix} x' & y' \end{pmatrix} (Q^T AQ) \begin{pmatrix} x' \\ y' \end{pmatrix} .$$

The 2×2 matrix $Q^T AQ$ gives the coefficients of the x', y' quadratic form, but their values depend on the rotation matrix Q. What rotation is needed? We want one which will produce coordinate axes along the principal axes of the curve. The algebraic signal for this is the disappearance of the term $2b'x'y'$ in the quadratic form, i.e. the coefficient b' needs to be zero. ***Hence, the new matrix of coefficients $Q^T AQ$ must be diagonal.*** If we define a 2×2 diagonal matrix D by

$$(8.14) \qquad D = \begin{pmatrix} \lambda_1 & 0 \\ 0 & \lambda_2 \end{pmatrix} ,$$

where λ_1 and λ_2 are as yet undetermined, then the rotation matrix Q ***must*** satisfy

$$Q^T A Q = D \quad . \tag{8.15}$$

Thus, Equation (8.15) will have the following desired form

$$-f = \mathbf{v}'^T D \mathbf{v}' = \begin{pmatrix} x' & y' \end{pmatrix} \begin{pmatrix} \lambda_1 & 0 \\ 0 & \lambda_2 \end{pmatrix} \begin{pmatrix} x' \\ y' \end{pmatrix} = \lambda_1 x'^2 + \lambda_2 y'^2 \quad . \tag{8.16}$$

Thus, starting from the symmetric matrix A obtained from the original Equation (8.9), we need to solve Equation (8.15) for matrices Q and D. But this is just the eigenvalue problem of Chapter 7 in disguise as we shall now show.

Theorem 8.1. *If the diagonal elements of the matrix D are the eigenvalues of the matrix A, and the columns of the matrix Q are chosen to be the corresponding unit length mutually orthogonal eigenvectors of A, then $Q^T A Q = D$ and Q is an orthogonal matrix.*

Proof. Since A, the matrix of coefficients of the original equation, is a symmetric matrix, Theorem 7.2 guarantees the existence of a complete set of mutually orthogonal eigenvectors. Let $\mathbf{q}_1$ and $\mathbf{q}_2$ be such a set with lengths adjusted to be 1, and let λ_1 and λ_2 be the corresponding eigenvalues. Then

$$A\mathbf{q}_1 = \lambda_1 \mathbf{q}_1 \quad \text{and} \quad A\mathbf{q}_2 = \lambda_2 \mathbf{q}_2 \quad . \tag{8.17}$$

Now choose $\mathbf{q}_1$ and $\mathbf{q}_2$ to be columns of Q. Since they are each of unit length and mutually orthogonal, we know that Q is an orthogonal matrix. Letting the diagonal elements of D be λ_1 and λ_2, careful thinking about matrix multiplication gives:

$$AQ = A \begin{pmatrix} \mathbf{q}_1 & \mathbf{q}_2 \end{pmatrix} = \begin{pmatrix} A\mathbf{q}_1 & A\mathbf{q}_2 \end{pmatrix} = \begin{pmatrix} \lambda_1 \mathbf{q}_1 & \lambda_2 \mathbf{q}_2 \end{pmatrix} =$$

$$\begin{pmatrix} \mathbf{q}_1 & \mathbf{q}_2 \end{pmatrix} \begin{pmatrix} \lambda_1 & 0 \\ 0 & \lambda_2 \end{pmatrix} = QD \quad . \tag{8.18}$$

Multiplying this equation from the left by Q^T, we then get $Q^T A Q = D$ as desired. ∎

8.2 Summary of the Method for Solving the Principal Axes Problem

Given an equation of the form (8.1):

1. Find the symmetric matrix A of the quadratic form in $\mathbf{v}$ for the equation.

2. Solve the eigenvalue problem for A.

3. Write the new equation in $\mathbf{v}'$. The eigenvalues of A are the coefficients in the new quadratic form in $\mathbf{v}'$.

4. Sketch the principal axes. The eigenvectors lie along the (*positive*) principal axes.

5. Find the matrix Q with $\det Q = 1$ whose columns are the eigenvectors of unit length. (*If you obtained an orthogonal matrix with determinant* -1 *you can multiply the first column by* -1 *to obtain a rotation matrix.*)

Example 8.1 Graph the curve described by the equation $xy = 1$.

Solution.

1) The quadratic form to be considered is $0x^2 + xy + 0y^2$, so

$$A = \begin{pmatrix} 0 & \frac{1}{2} \\ \frac{1}{2} & 0 \end{pmatrix} . \tag{8.19}$$

2) Solving the eigenvalue problem for this A, we find the characteristic equation to be given by

$$\lambda^2 - \frac{1}{4} = 0 \quad , \tag{8.20}$$

so the eigenvalues are $\lambda = \pm\frac{1}{2}$. By the methods of Chapter 7, corresponding pairs of eigenvalues and eigenvectors are found to be

$$\lambda_1 = \frac{1}{2}, \quad \mathbf{v}_1 = \begin{pmatrix} 1 \\ 1 \end{pmatrix} \quad \text{and} \quad \lambda_2 = \frac{-1}{2}, \quad \mathbf{v}_2 = \begin{pmatrix} 1 \\ -1 \end{pmatrix} . \tag{8.21}$$

3) The order in which we label the eigenvalues is arbitrary. But once chosen, the new quadratic form is determined by (8.16). Hence with this labeling we have

$$\begin{pmatrix} x' & y' \end{pmatrix} \begin{pmatrix} \frac{1}{2} & 0 \\ 0 & \frac{-1}{2} \end{pmatrix} \begin{pmatrix} x' \\ y' \end{pmatrix} = \frac{1}{2}x'^2 - \frac{1}{2}y'^2 . \tag{8.22}$$

Thus the new equation for the curve is

$$\frac{1}{2}x'^2 - \frac{1}{2}y'^2 = 1 \tag{8.23}$$

which is the standard form for a horizontally oriented hyperbola with center at the origin and asymptotes $y' = \pm x'$.

4) The directions of the new x', y' axes are those of the eigenvectors, $\mathbf{v}_1$ and $\mathbf{v}_2$. There is still the choice of which way along each new axis to take as the positive direction, since $-\mathbf{v}_1$ is just as valid a choice as $\mathbf{v}_1$ for an eigenvector. We normally choose positive directions so that the new axes form a right-handed system. Then the two coordinate systems are related by a rotation matrix Q since $\det Q = 1$. Here we choose $\mathbf{v}_1$ and $-\mathbf{v}_2$ to determine the positive directions along the new coordinate axes. See Figure 8.1.

5) For this problem we do not have to find Q, but will give it for illustration purposes. Since the lengths of $\mathbf{v}_1$ and $-\mathbf{v}_2$ are each $\sqrt{2}$, the corresponding columns of Q will be $\mathbf{q}_1 = (1/\sqrt{2})\mathbf{v}_1$ and $\mathbf{q}_2 = -(1/\sqrt{2})\mathbf{v}_2$. Thus

$$Q = \frac{1}{\sqrt{2}}\begin{pmatrix} 1 & -1 \\ 1 & 1 \end{pmatrix} \quad . \tag{8.24}$$

8.3 The General Second Degree Equation in Two Variables

The general second degree equation in two variables x, y is an extension of Equation (8.1) to include first degree terms:

$$ax^2 + 2bxy + cy^2 + dx + ey + f = 0 \quad , \tag{8.25}$$

where a, b, c, d, e, and f are constants. The goal is again to obtain the simplest equation for the curve by making a better choice of coordinates than x and y seem to be. The curve will then be most easily graphed relative to the preferred axes.

The method is a simple extension of the one we have just discussed. Start by finding the symmetric matrix A from the coefficients of the second degree terms to obtain

$$A = \begin{pmatrix} a & b \\ b & c \end{pmatrix} \quad . \tag{8.26}$$

Use its eigenvalues and eigenvectors to discover the new axes x', y', and to find an appropriate rotation matrix Q. Then we have

$$\mathbf{v} = Q\mathbf{v}' \quad . \tag{8.27}$$

Now Equation (8.25) can be expressed in matrix form by

$$\begin{pmatrix} x & y \end{pmatrix}\begin{pmatrix} a & b \\ b & c \end{pmatrix}\begin{pmatrix} x \\ y \end{pmatrix} + \begin{pmatrix} d & e \end{pmatrix}\begin{pmatrix} x \\ y \end{pmatrix} = -f$$

or

(8.28) $$\mathbf{v}^T A\mathbf{v} + \mathbf{p}^T\mathbf{v} = -f$$

where $\mathbf{p}^T = \begin{pmatrix} d & e \end{pmatrix}$. Since $\mathbf{v} = Q\mathbf{v}'$ and $Q^T AQ = D$, Equation (8.28) becomes:

(8.29) $$\mathbf{v}'^T D\mathbf{v}' + \mathbf{p}'^T\mathbf{v}' = -f \quad .$$

where $\mathbf{p}'^T = \mathbf{p}^T Q$ or $\mathbf{p}' = Q^T\mathbf{p}$. Equation (8.29) is Equation (8.25) expressed completely in terms of the new variables, the principal axes variables. Namely

(8.30) $$\lambda_1 x'^2 + \lambda_2 y'^2 + d'x' + e'y' + f' = 0$$

where $\begin{pmatrix} d' & e' \end{pmatrix} = \mathbf{p}'^T$ and $f' = f$.

Example 8.2 Graph the equation

(8.31) $$xy + 2\sqrt{2}\,x - \sqrt{2}\,y = 1 \quad .$$

Solution.

The matrix A is just the one from Equation (8.19), so we know the principal axes and the rotation matrix Q from Equation (8.24). Since $d = 2\sqrt{2}$, $e = -\sqrt{2}$, and $f = -1$ from (8.31), we may use Equation (8.29) to form the curve's equation in terms of the new variables. Thus,

(8.32) $$\frac{1}{2}x'^2 - \frac{1}{2}y'^2 + \begin{pmatrix} 2\sqrt{2} & -\sqrt{2} \end{pmatrix}\begin{pmatrix} \frac{1}{\sqrt{2}} & \frac{-1}{\sqrt{2}} \\ \frac{1}{\sqrt{2}} & \frac{1}{\sqrt{2}} \end{pmatrix}\begin{pmatrix} x' \\ y' \end{pmatrix} = 1 \quad .$$

Performing the matrix multiplication gives us

(8.33) $$\frac{1}{2}x'^2 - \frac{1}{2}y'^2 + x' - 3y' = 1 \quad .$$

Apply the technique of completing the square to each variable to obtain

(8.34) $$(x'+1)^2 - (y'+3)^2 = -6 \quad .$$

Thus we needed a translation of the x', y' axes to obtain the center $(h_1', h_2') = (-1, -3)$ of the conic, a hyperbola. See Figure 8.2.

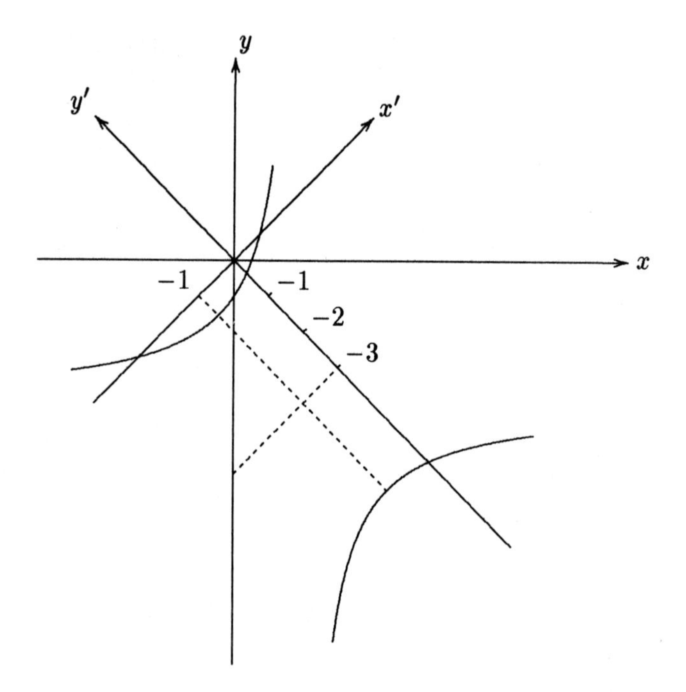

Figure 8.2

8.4 The General Second Degree Equation in Three Variables

A major advantage of matrix equations is their independence of the dimension of the space. The methods we have developed for two variables when written in matrix language hold for three and higher dimensions as well. For three variables x_1, x_2, and x_3 the general second degree equation is given by

$$a_{11}x_1^2 + a_{22}x_2^2 + a_{33}x_3^2 + 2a_{12}x_1x_2 + 2a_{13}x_1x_3 + 2a_{23}x_2x_3 \tag{8.35}$$

$$+b_1x_1 + b_2x_2 + b_3x_3 + c = 0$$

where a_{ij}, $i,j = 1,2,3$, b_k, $k = 1,2,3$, and c are constants and the variables x_1, x_2, and x_3 have been subscripted so that it is easier to keep track of the terms in Equation (8.35). If A is the symmetric matrix

$$A = \begin{pmatrix} a_{11} & a_{12} & a_{13} \\ a_{21} & a_{22} & a_{23} \\ a_{31} & a_{32} & a_{33} \end{pmatrix} , \tag{8.36}$$

if $\mathbf{b}$ is the column vector

$$\mathbf{b} = \begin{pmatrix} b_1 \\ b_2 \\ b_3 \end{pmatrix} , \tag{8.37}$$

and if $\mathbf{v}$ is the coordinate column

$$\mathbf{v} = \begin{pmatrix} x_1 \\ x_2 \\ x_3 \end{pmatrix} , \tag{8.38}$$

then Equation (8.35) can be expressed in matrix form by

$$\begin{pmatrix} x_1 & x_2 & x_3 \end{pmatrix} \begin{pmatrix} a_{11} & a_{12} & a_{13} \\ a_{21} & a_{22} & a_{23} \\ a_{31} & a_{32} & a_{33} \end{pmatrix} \begin{pmatrix} x_1 \\ x_2 \\ x_3 \end{pmatrix} \tag{8.39}$$

$$+ \begin{pmatrix} b_1 & b_2 & b_3 \end{pmatrix} \begin{pmatrix} x_1 \\ x_2 \\ x_3 \end{pmatrix} + c = 0 ,$$

or more compactly by

$$\mathbf{v}^T A\mathbf{v} + \mathbf{b}^T\mathbf{v} + c = 0 \quad . \tag{8.40}$$

Here A is the matrix of the quadratic part of the equation. In order to analyze the graph of (8.40) we need to find the eigenvalues λ_1, λ_2, and λ_3 of the symmetric matrix A and corresponding orthonormal eigenvectors $\mathbf{q}_1$, $\mathbf{q}_2$, and $\mathbf{q}_3$. (*Choosing them so that they form a right handed set of vectors*). Then the matrix $Q = [\mathbf{q}_1, \mathbf{q}_2, \mathbf{q}_3]$ is a rotation of coordinate axes determined by $\{\mathbf{q}_1, \mathbf{q}_2, \mathbf{q}_3\}$. Let

$$\mathbf{v}' = \begin{pmatrix} x_1' \\ x_2' \\ x_3' \end{pmatrix} \tag{8.41}$$

be the new coordinate column and let

$$D = \begin{pmatrix} \lambda_1 & 0 & 0 \\ 0 & \lambda_2 & 0 \\ 0 & 0 & \lambda_3 \end{pmatrix} \tag{8.42}$$

be the diagonal matrix of eigenvalues of A. Then $\mathbf{v} = Q\mathbf{v}'$ and Equation (8.40) is transformed into

$$\begin{aligned} 0 &= \mathbf{v}^T A\mathbf{v} + \mathbf{b}^T\mathbf{v} + c = (Q\mathbf{v}')^T A(Q\mathbf{v}') + \mathbf{b}^T(Q\mathbf{v}') + c \\ &= \mathbf{v}'^T(Q^T AQ)\mathbf{v}' + (\mathbf{b}^T Q)\mathbf{v}' + c \\ &= \mathbf{v}'^T D\mathbf{v}' + (Q^T\mathbf{b})^T\mathbf{v}' + c \\ &= \mathbf{v}'^T D\mathbf{v}' + \mathbf{b}'^T\mathbf{v}' + c \end{aligned} \tag{8.43}$$

where

$$\mathbf{b}' = Q^T\mathbf{b} \tag{8.44}$$

is the new coordinate column corresponding to the old coordinate column $\mathbf{b}$. In terms of matrices and coordinates Equation (8.43) can now be expressed by

$$\lambda_1 x_1'^2 + \lambda_2 x_2'^2 + \lambda_3 x_3'^2 + b_1' x_1' + b_2' x_2' + b_3' x_3' + c = 0 \quad . \tag{8.45}$$

Then, to finish the analysis of the equation, you complete the square to find a translation of the origin to a point with new primed coordinates (h_1', h_2', h_3') whose new coordinate column $\mathbf{h}$ is given by

$$\mathbf{h}' = \begin{pmatrix} h_1' \\ h_2' \\ h_3' \end{pmatrix} \quad . \tag{8.46}$$

Then if

$$\mathbf{v}'' = \begin{pmatrix} x_1'' \\ x_2'' \\ x_3'' \end{pmatrix} \tag{8.47}$$

is the coordinate column for the new coordinates after the translation of axes we have

$$\mathbf{v}'' = \mathbf{v}' - \mathbf{h}' \tag{8.48}$$

or

$$\mathbf{v}' = \mathbf{v}'' + \mathbf{h}' \quad . \tag{8.49}$$

Then, Equation (8.45) transforms into

$$\lambda_1 x_1''^2 + \lambda_2 x_2''^2 + \lambda_3 x_3''^2 + \tilde{c} = 0 \tag{8.50}$$

or

$$\lambda_1(x_1' - h_1')^2 + \lambda_2(x_2' - h_2')^2 + \lambda_3(x_3' - h_3')^2 + \tilde{c} = 0 \quad . \tag{8.51}$$

From Equation (8.51) we can identify the standard quadric surface corresponding to the original Equation (8.35). In the primed (*rotated*) coordinate system the coordinates of the center of this quadratic surface are given by $\mathbf{h}'$ and we can use $\mathbf{h} = Q\mathbf{h}'$ to find the coordinates of the center in the original coordinates.

We will demonstrate the three dimensional case with the following example.

Example 8.3 Analyze the graph of the equation

$$5x^2 + 5y^2 + 8z^2 - 4xy + 2xz - 2yz + 3\sqrt{2}\,x + 3\sqrt{2}\,y + 12\sqrt{3}\,z + 14 = 0 \quad . \tag{8.52}$$

Solution.

The matrix A of the equation is

(8.53)
$$A = \begin{pmatrix} 5 & -2 & 1 \\ -2 & 5 & -1 \\ 1 & -1 & 8 \end{pmatrix} ,$$

where each diagonal element is the coefficient of the appropriate squared term in Equation (8.52) and each off-diagonal element is half the coefficient of the appropriate product term. Its eigenvalue and eigenvector pairs are found to be given by

(8.54)
$$\left\{3, \begin{pmatrix} 1 \\ 1 \\ 0 \end{pmatrix}\right\} \left\{6, \begin{pmatrix} -1 \\ 1 \\ 1 \end{pmatrix}\right\} \left\{9, \begin{pmatrix} 1 \\ -1 \\ 2 \end{pmatrix}\right\} .$$

The matrix Q with $\det Q = 1$ (*whose columns are composed of eigenvectors of unit length*) is given by

(8.55)
$$Q = \begin{pmatrix} \frac{1}{\sqrt{2}} & \frac{-1}{\sqrt{3}} & \frac{1}{\sqrt{6}} \\ \frac{1}{\sqrt{2}} & \frac{1}{\sqrt{3}} & \frac{-1}{\sqrt{6}} \\ 0 & \frac{1}{\sqrt{3}} & \frac{2}{\sqrt{6}} \end{pmatrix} .$$

Then since $\mathbf{v} = Q\mathbf{v}'$ Equation (8.52) becomes

$$3x'^2 + 6y'^2 + 9z'^2 + \begin{pmatrix} 3\sqrt{2} & 3\sqrt{2} & 12\sqrt{3} \end{pmatrix} Q \begin{pmatrix} x' \\ y' \\ z' \end{pmatrix} + 14 = 0 ,$$

which is

(8.56)
$$3x'^2 + 6y'^2 + 9z'^2 + 6x' + 12y' + 12\sqrt{2}\, z' = -14 .$$

After completing the square in each variable, we obtain

$$3(x'+1)^2 + 6(y'+1)^2 + 9\left(z' + \frac{2\sqrt{2}}{3}\right)^2 = 3 .$$

Dividing by 3, the equation becomes:

(8.57)
$$(x'+1)^2 + 2(y'+1)^2 + 3\left(z' + \frac{2\sqrt{2}}{3}\right)^2 = 1 .$$

Thus the graph is an ellipsoid centered at the point $h_1' = -1$, $h_2' = -1$, $h_3' = -2\sqrt{2}/3$, or correspondingly (*using* $\mathbf{v} = Q\mathbf{v}'$), $h_1 \approx -0.515$, $h_2 \approx -0.900$, $h_3 \approx -1.347$. The ellipsoid's principal axes are in the directions of the eigenvectors of A given in Equation (8.54).

8.5 Exercises

1. Find the symmetric matrix A for each of the following equations

a) $13x^2 + 19y^2 - 12xy = 5$

b) $\begin{pmatrix} x & y \end{pmatrix} \begin{pmatrix} 3 & -5 \\ 7 & 9 \end{pmatrix} \begin{pmatrix} x \\ y \end{pmatrix} + \begin{pmatrix} x & y \end{pmatrix} \begin{pmatrix} 3x+2y \\ -2x+5y \end{pmatrix} = 24$

c) $\begin{pmatrix} x & y \end{pmatrix} \begin{pmatrix} 3 & -5 \\ 9 & 7 \end{pmatrix} \begin{pmatrix} x \\ y \end{pmatrix} + \begin{pmatrix} -2x+3y & 7x-2y \end{pmatrix} \begin{pmatrix} x \\ y \end{pmatrix}$
$+ \begin{pmatrix} x & y \end{pmatrix} \begin{pmatrix} 3x+y \\ 3x-y \end{pmatrix} = -12$.

2. For each of the following symmetric matrices A with given eigenvalues λ_1 and λ_2 find the corresponding unit eigenvectors $\mathbf{q}_1$ and $\mathbf{q}_2$ corresponding to a rotation of coordinate axes. Then express the quadratic equation $\mathbf{v}^T A\mathbf{v} = 1$ in terms of the new coordinates (x', y') corresponding to the rotated coordinate axes.

a) $A = \begin{pmatrix} -4 & 6 \\ 6 & 1 \end{pmatrix}$, $\lambda_1 = 5$ and $\lambda_2 = -8$

b) $A = \begin{pmatrix} 5 & -10 \\ -10 & -16 \end{pmatrix}$, $\lambda_1 = 9$ and $\lambda_2 = -20$

c) $A = \begin{pmatrix} -3 & 12 \\ 12 & 4 \end{pmatrix}$, $\lambda_1 = 13$, and $\lambda_2 = -12$.

3. For each of the following equations, solve the principal axes problem (find the *principal axes and sketch the graph*).

a) $3x^2 + 2xy + 3y^2 = 16$

b) $x^2 + 4xy - 2y^2 = 6$

c) $2x^2 + 2xy + 2y^2 = \sqrt{2}\,x - \sqrt{2}\,y$.

4. What can you say about the eigenvalues of the symmetric matrix A if the graph of the equation

$$\begin{pmatrix} x & y \end{pmatrix} A \begin{pmatrix} x \\ y \end{pmatrix} + \begin{pmatrix} d & e \end{pmatrix} \begin{pmatrix} x \\ y \end{pmatrix} + f = 0$$

is

a) an ellipse **b)** a hyperbola **c)** a parabola?

5. Find the symmetric matrix A for each of the following equations:

a) $13x^2 - 24xy + 8xz - 29y^2 + 34yz + 41z^2 = 12$

b) $\begin{pmatrix} x & y & z \end{pmatrix} \begin{pmatrix} 2 & -8 & 16 \\ 0 & 5 & 28 \\ 0 & 0 & 3 \end{pmatrix} \begin{pmatrix} x \\ y \\ z \end{pmatrix} = 56$

c) $\begin{pmatrix} x & y & z \end{pmatrix} \begin{pmatrix} 2 & 8 & 26 \\ -2 & 7 & 15 \\ 4 & 3 & 9 \end{pmatrix} \begin{pmatrix} x \\ y \\ z \end{pmatrix} + \begin{pmatrix} x & y & z \end{pmatrix} \begin{pmatrix} x - 6y + z \\ 2x + y + 2z \\ x + 6y + z \end{pmatrix} = 22$

d) $16y^2 + 34yz + 9x^2 + 22xz + 14xy + 25z^2 = 15$.

6. The eigenvalues of

$$A = \begin{pmatrix} 5 & -1 & -1 \\ -1 & 1 & 3 \\ -1 & 3 & 1 \end{pmatrix}$$

are $\lambda_1 = 3$, $\lambda_2 = 6$, and $\lambda_3 = -2$.

a) Find corresponding unit eigenvectors.

b) Find a rotation matrix Q so that $Q^T A Q = D$ is in diagonal form.

c) Express the equation $\mathbf{v}^T A \mathbf{v} = 100$ in terms of the new coordinates $\mathbf{v}' = [x', y', z']^T$ corresponding to Q.

7. Repeat Problem 6 for the matrix

$$A = \begin{pmatrix} 1 & 1 & -4 \\ 1 & 4 & -7 \\ -4 & -7 & 9 \end{pmatrix}$$

with eigenvalues $\lambda_1 = 1$, $\lambda_2 = -2$, and $\lambda_3 = 15$.

8. We know that $\lambda_1 = 9$ is an eigenvalue of the matrix

$$A = \begin{pmatrix} 10 & -3 & -2 \\ -3 & 7 & -5 \\ -2 & -5 & 2 \end{pmatrix}$$

and that

$$\mathbf{p}_2 = \begin{pmatrix} 1 \\ 2 \\ 3 \end{pmatrix}$$

is an eigenvector corresponding to another eigenvalue λ_2 of A.

a) Find an eigenvector $\mathbf{p}_1$ corresponding to $\lambda_1 = 9$

b) Find the eigenvalues λ_2 and λ_3

c) Find unit eigenvectors $\mathbf{q}_1$, $\mathbf{q}_2$, and $\mathbf{q}_3$ for A such that $Q = [\mathbf{q}_1, \mathbf{q}_2, \mathbf{q}_3]$ is a rotation matrix

d) Express the equation $\mathbf{v}^T A\mathbf{v} = 1$ in terms of the new coordinates $\mathbf{v}' = [x', y', z']^T$ corresponding to Q.

9. Repeat Problem 8 if A is the matrix

$$A = \begin{pmatrix} -1 & 6 & -4 \\ 6 & 2 & -2 \\ -4 & -2 & 7 \end{pmatrix}$$

$\lambda_1 = 11$ is an eigenvalue of A and

$$\mathbf{p}_2 = \begin{pmatrix} 1 \\ 2 \\ 2 \end{pmatrix}$$

is an eigenvector corresponding to another eigenvalue λ_2 of A.

10. Let $\{\mathbf{p}_1, \mathbf{p}_2, \mathbf{p}_3\}$ be mutually orthogonal 3×1 column vectors with $|\mathbf{p}_1| = 1$, $|\mathbf{p}_2| = 2$, $|\mathbf{p}_3| = 3$, and $\det[\mathbf{p}_1, \mathbf{p}_2, \mathbf{p}_3] = -6$. Let λ_1, λ_2, λ_3 be the eigenvalues of a symmetric matrix A which has the eigenvectors $\mathbf{p}_1$, $\mathbf{p}_2$, and $\mathbf{p}_3$ respectively. Explain how to construct a rotation matrix Q such that an equation $\mathbf{v}^T A\mathbf{v} = 1$ transforms into the equation

$$\mathbf{v}'^T Q^T AQ\mathbf{v}' = \lambda_2 x'^2 + \lambda_1 y'^2 + \lambda_3 z'^2 = 1$$

in terms of the new x', y', z' coordinates corresponding to Q.

11. The matrix

$$A = \begin{pmatrix} -5 & 3 & -5 \\ 3 & 3 & -15 \\ -5 & -15 & 19 \end{pmatrix}$$

has $\lambda_1 = \lambda_2 = -6$ as a twice repeated eigenvalue.

a) Find an eigenvector $\mathbf{p}_3$ corresponding to the third eigenvalue λ_3. Then use direct matrix multiplication to find λ_3 and verify that $A\mathbf{p}_3 = \lambda_3\mathbf{p}_3$. (Do not *compute the characteristic equation for A and do not use the value of* λ_3 *to find* $\mathbf{p}_3$).

b) The vector $\mathbf{p}_1 = [-1, 2, 1]^T$ is one of the eigenvectors of A corresponding to $\lambda_1 = -6$. Find a second eigenvector $\mathbf{p}_2$ corresponding to $\lambda_2 = -6$ so that $\{\mathbf{q}_1, \mathbf{q}_2, \mathbf{q}_3\}$ is a right handed set of vectors where $\mathbf{q}_k = \mathbf{p}_k/|\mathbf{p}_k|$ for $k = 1, 2, 3$.

c) Transform the equation $\mathbf{v}^T A\mathbf{v} = 1$ into a new equation in terms of the new x', y', z' coordinates corresponding to Q.

12. **(a)** Find the unit vectors along the principal axes and **(b)** Identify the quadric surface described by

$$7x^2 + 7y^2 + 10z^2 - 2xy - 4xz + 4yz = 24 \quad .$$

13. Suppose that

$$x^2 - 6y^2 + 8z^2 + 8xy + 8xz - 3x + 6z = 48 \quad .$$

a) Find three orthogonal unit vectors along the principal axes for the quadratic part of the above equation.

b) Let x', y', and z' be the new coordinates corresponding to the principal axes of the above equation. Transform the above equation into these new coordinates.

14. **a)** Use matrix multiplication to show that

$$\mathbf{w}_1 = \begin{pmatrix} 2 \\ -1 \\ 2 \end{pmatrix}, \quad \mathbf{w}_2 = \begin{pmatrix} -1 \\ 2 \\ 2 \end{pmatrix}, \quad \text{and} \quad \mathbf{w}_3 = \begin{pmatrix} -2 \\ -2 \\ 1 \end{pmatrix} ,$$

are eigenvectors of the matrices A and B given by

$$A = \begin{pmatrix} -1 & 2 & -2 \\ 2 & 0 & 0 \\ -2 & 0 & -2 \end{pmatrix}$$

and

$$B = \begin{pmatrix} 1 & -2 & 0 \\ -2 & 2 & 2 \\ 0 & 2 & 3 \end{pmatrix} ,$$

and to find the corresponding ***eigenvalues*** of A and B.

b) Using the matrices A and B from part **(a)** determine the ***standard forms*** of the quadratic surfaces

$$\begin{pmatrix} x & y & z \end{pmatrix} A \begin{pmatrix} x \\ y \\ z \end{pmatrix} = 40 ,$$

$$\begin{pmatrix} x & y & z \end{pmatrix} B \begin{pmatrix} x \\ y \\ z \end{pmatrix} = 40 ,$$

and

$$\begin{pmatrix} x & y & z \end{pmatrix} B \begin{pmatrix} x \\ y \\ z \end{pmatrix} = 0$$

in the new rotated (x', y', z')–coordinate system whose positive coordinate axes are given by $\{\mathbf{w}_1, \mathbf{w}_2, \mathbf{w}_3\}$.

c) Give ***sketches*** of the surfaces in **b)** with respect to the (x', y', z')–coordinate system and ***name*** the surfaces.

15. Transform the equation

$$-5x^2 - 6y^2 - 4z^2 + 4xy + 4xz + x - y + z = -26$$

into an equation involving the new coordinates x', y', and z' corresponding to the principal axes of the quadratic part of this equation.

9 Miscellaneous Problems for Chapters 1-8

1. Evaluate $\begin{pmatrix} -1 & 1 \\ 2 & 1 \\ 1 & -1 \end{pmatrix}^T \begin{pmatrix} 1 & 1 \\ -1 & 1 \\ 2 & 2 \end{pmatrix}$.

2. Let $Q = \frac{1}{9}\begin{pmatrix} 8 & -1 & -4 \\ 4 & 4 & 7 \\ -1 & 8 & -4 \end{pmatrix}$ represent the change of coordinates from (x, y, z) to (x', y', z').

a) Let $x = 3$, $y = -2$, and $z = -2$. Find x', y', and z'.

b) Let $x' = 1$, $y' = -2$, and $z' = -2$. Find x, y, and z.

3. Suppose that $AX = B$, $A = \begin{pmatrix} 3 & 4 \\ 2 & 3 \end{pmatrix}$, and $B = \begin{pmatrix} -1 & 2 & -2 \\ 1 & 1 & 3 \end{pmatrix}$. Find X.

4. Find all solutions of

$$\begin{pmatrix} -2 & 4 & -1 & -8 \\ -2 & 4 & 0 & -6 \\ 1 & -2 & 1 & 5 \end{pmatrix} X = \begin{pmatrix} 1 \\ 1 \\ 1 \end{pmatrix}.$$

5. Suppose that $A^{-1}X = B$ where $A = \begin{pmatrix} 3 & -2 & 4 \\ 4 & 5 & 2 \\ -5 & 2 & 7 \end{pmatrix}$ and $B = \begin{pmatrix} 1 & 1 \\ -1 & -1 \\ 1 & -1 \end{pmatrix}$. Find X.

6. Let $A = \begin{pmatrix} 1 & 2 \\ 3 & 6 \end{pmatrix}$, $B = \begin{pmatrix} 1 & -3 & 4 \\ 1 & 5 & 3 \end{pmatrix}$, and $C = \begin{pmatrix} -2 & 4 \\ 1 & -2 \end{pmatrix}$. Compute the following or state that the dimensions are not compatible.

a) AC **b)** CA **c)** C^TB **d)** $B + C$ **e)** $2A + C$ **f)** AB^T .

7. Find the volume of the parallelepiped with edge vectors $\mathbf{a} = \langle 2, 1, 1\rangle$, $\mathbf{b} = \langle 1, 1, -1\rangle$, and $\mathbf{c} = \langle 1, 1, 1\rangle$.

8. Suppose that $\begin{pmatrix} 6 & 0 & 1 \\ 2 & 3 & 1 \\ 3 & 2 & 1 \end{pmatrix} X = \begin{pmatrix} 1 & 0 & 0 \\ 0 & 1 & 0 \\ 0 & 0 & 1 \end{pmatrix}$. Find X.

9. Find the determinant of the following matrices

$$\text{a)}\ \begin{pmatrix} 1 & 1 & 1 & 1 \\ 1 & 2 & 2 & 2 \\ 1 & 2 & 3 & 3 \\ 1 & 2 & 3 & 4 \end{pmatrix} \quad \text{b)}\ \begin{pmatrix} 1 & 1 & -1 & 1 \\ -1 & 2 & 1 & -1 \\ 1 & -1 & 1 & 2 \\ 2 & 1 & 2 & 1 \end{pmatrix} \quad \text{c)}\ \begin{pmatrix} 1 & -1 & 2 & 1 \\ 0 & 1 & -1 & 1 \\ -1 & 3 & 1 & 2 \\ 2 & 1 & 3 & 3 \end{pmatrix}$$

$$\text{d)}\ \begin{pmatrix} 1 & 0 & -1 & 2 \\ 2 & 1 & 3 & -1 \\ -1 & 1 & 2 & 1 \\ 1 & 3 & -1 & 2 \end{pmatrix} \quad \text{e)}\ \begin{pmatrix} 1 & -1 & 2 & 1 \\ -1 & 1 & 1 & -1 \\ 3 & -1 & 2 & 1 \\ 4 & 1 & 1 & 3 \end{pmatrix}.$$

10. Find all vectors X that satisfy the matrix equation

$$\begin{pmatrix} 0 & 1 & -3 & 0 & 0 & 2 \\ 0 & 0 & 0 & 1 & 0 & -5 \\ 0 & 0 & 0 & 0 & 1 & 4 \\ 0 & 0 & 0 & 0 & 0 & 0 \end{pmatrix} X = \begin{pmatrix} 8 \\ 6 \\ 7 \\ 0 \end{pmatrix}.$$

11. Find the inverse of $\begin{pmatrix} 3 & 2 & 1 \\ 2 & 2 & 1 \\ 1 & 1 & 1 \end{pmatrix}$.

12. Find all solutions of

$$\begin{pmatrix} 1 & -3 & 3 \\ 2 & -7 & 8 \\ 3 & -11 & 14 \\ 1 & -5 & 9 \end{pmatrix} X = \begin{pmatrix} 21 \\ 52 \\ 87 \\ 49 \end{pmatrix}.$$

13. Write the coefficient matrix and the augmented matrix of the following systems of equations.

$$\text{a)}\quad x + 2y - z = 3 \qquad \text{b)}\quad \begin{aligned} x_1 + 3x_2 + 5x_3 &= 7 \\ 2x_1 + 4x_2 + 6x_3 &= 8 \end{aligned}$$

$$\text{c)}\quad \begin{aligned} x + y &= 0 \\ 2x + y &= 1 \\ 3x + y &= 2 \end{aligned} \qquad \text{d)}\quad \begin{aligned} x + y + z &= 0 \\ x - y + z &= 1 \\ x - y - z &= 2 \end{aligned}.$$

14. Each of the following $m \times n$ matrices is an augmented matrix of a system of equations in variables $x_1, x_2, \ldots x_{n-1}$. Write out the system of equations for each of the following matrices.

a) $\left(\begin{array}{c|c} 2 & 3 \end{array}\right)$ **b)** $\left(\begin{array}{cc|c} 2 & 1 & 0 \\ 3 & 4 & 5 \end{array}\right)$ **c)** $\left(\begin{array}{cc|c} 1 & 3 & 1 \\ 2 & 1 & 2 \\ 3 & 2 & 1 \end{array}\right)$.

15. Solve the following systems of equations

a) $\begin{aligned} 2x + 3y &= 5 \\ 3x + 2y &= 5 \end{aligned}$ **b)** $\begin{aligned} 2x + 3y &= 0 \\ 3x + 2y &= 5 \end{aligned}$ **c)** $\begin{aligned} 2x + 3y &= 5 \\ 3x + 2y &= 0 \end{aligned}$

d) $\begin{aligned} 2x + 3y &= 0 \\ 3x + 2y &= 0 \end{aligned}$.

16. Solve the following systems of equations

a) $\begin{aligned} x + 2y &= -1 \\ x + 3y &= -3 \\ 2x + 5y &= -4 \end{aligned}$ **b)** $\begin{aligned} x + 2y &= 2 \\ x + 3y &= 1 \\ 2x + 5y &= 0 \end{aligned}$ **c)** $\begin{aligned} x + 2y &= 0 \\ x + 3y &= 0 \\ 2x + 5y &= 0 \end{aligned}$

d) $\begin{aligned} x + 2y &= 1 \\ x + 3y &= 2 \\ 2x + y &= 3 \end{aligned}$.

17. Solve the following systems of equations

a) $\begin{aligned} 2x_1 + x_2 - 4x_3 + 4x_4 &= 1 \\ x_1 + 2x_2 + x_3 + 2x_4 &= 2 \end{aligned}$ **b)** $\begin{aligned} x_1 + x_2 - x_3 - x_4 &= 0 \\ x_2 - x_3 + x_4 &= 0 \\ x_3 + x_4 &= 0 \end{aligned}$

c) $\begin{aligned} x_1 + 2x_2 - x_3 &= 2 \\ 2x_1 - x_2 + 3x_3 &= 4 \\ 3x_1 + x_2 + 2x_3 &= 6 \end{aligned}$ **d)** $\begin{aligned} x_1 + 2x_2 + x_3 + x_4 &= 0 \\ 2x_1 + x_2 + 3x_3 - x_4 &= -4 \\ x_1 - x_2 + 0x_3 + x_4 &= 7 \\ 3x_1 + 3x_2 + 7x_3 + x_4 &= -4 \end{aligned}$.

18. Use elementary row operations to find row-reduced echelon matrices which are row equivalent to the following matrices.

a) $\begin{pmatrix} 2 & 2 & 1 & 1 \\ 1 & 1 & 1 & 1 \end{pmatrix}$ **b)** $\begin{pmatrix} 1 & 3 & 1 & -1 & 1 & 2 \\ 2 & 6 & -1 & 4 & 1 & 2 \\ 1 & 3 & 2 & -3 & 1 & 3 \end{pmatrix}$

c) $\begin{pmatrix} 1 & 1 & 3 & -2 & 1 \\ 2 & 1 & 4 & -3 & 5 \\ 3 & -1 & 1 & -2 & 2 \end{pmatrix}$ **d)** $\begin{pmatrix} 1 & 1 & 2 \\ 2 & 1 & 3 \\ 3 & 1 & 4 \\ 1 & 2 & 3 \end{pmatrix}$.

19. Evaluate the following determinants

a) $\begin{vmatrix} 3 & 2 \\ 5 & 1 \end{vmatrix}$ **b)** $\begin{vmatrix} 3 & -2 \\ 5 & 1 \end{vmatrix}$ **c)** $\begin{vmatrix} -3 & 2 \\ 5 & 1 \end{vmatrix}$ **d)** $\begin{vmatrix} -3 & 2 \\ 5 & -1 \end{vmatrix}$

e) $\begin{vmatrix} 3 & -2 \\ -5 & -1 \end{vmatrix}$.

20. Find A^{-1} and check your answers if A equals

a) $\begin{pmatrix} 3 & 2 \\ 4 & 5 \end{pmatrix}$ **b)** $\begin{pmatrix} 1 & 3 & 3 \\ 2 & 7 & 7 \\ 1 & 8 & 9 \end{pmatrix}$ **c)** $\begin{pmatrix} 1 & 5 & 2 & 1 \\ 0 & 1 & -2 & -3 \\ 0 & 0 & 1 & 2 \\ 0 & 0 & 0 & 1 \end{pmatrix}$

d) $\begin{pmatrix} 1 & -8 & 2 & 0 \\ -2 & 1 & 0 & 3 \\ 0 & -4 & 1 & 0 \\ 0 & 0 & 0 & 1 \end{pmatrix}$.

21. Solve the following equations

a) $\begin{pmatrix} 3 & 4 \\ 5 & 6 \end{pmatrix} \begin{pmatrix} x_{11} & x_{12} \\ x_{21} & x_{22} \end{pmatrix} = \begin{pmatrix} 1 & 3 \\ -2 & 1 \end{pmatrix}$

b) $\begin{pmatrix} 1 & 2 & 3 \\ -3 & 1 & 0 \\ 2 & 0 & 1 \end{pmatrix} \begin{pmatrix} x_{11} & x_{12} \\ x_{21} & x_{22} \\ x_{31} & x_{32} \end{pmatrix} = \begin{pmatrix} 1 & 7 \\ 1 & 11 \\ -1 & -3 \end{pmatrix}$

c) $\begin{pmatrix} 2 & 5 \\ 3 & 4 \end{pmatrix} \begin{pmatrix} x_{11} & x_{12} & x_{13} \\ x_{21} & x_{22} & x_{23} \end{pmatrix} = \begin{pmatrix} 1 & 0 & 2 \\ 0 & 1 & 3 \end{pmatrix}$

d) $\begin{pmatrix} 1 & 0 & 3 & 4 \\ 0 & 1 & 2 & 1 \\ 0 & 0 & 1 & 0 \\ 0 & 0 & 0 & 1 \end{pmatrix} \begin{pmatrix} x_{11} & x_{12} \\ x_{21} & x_{22} \\ x_{31} & x_{32} \\ x_{41} & x_{42} \end{pmatrix} = \begin{pmatrix} 1 & 1 \\ -1 & 6 \\ 1 & 2 \\ -1 & 1 \end{pmatrix}$.

22. Solve the following systems of equations and also find the inverse of each coefficient matrix

a) $\begin{pmatrix} 2 & -1 \\ 1 & 3 \end{pmatrix} \begin{pmatrix} x_1 \\ x_2 \end{pmatrix} = \begin{pmatrix} 1 \\ 2 \end{pmatrix}$ **b)** $\begin{pmatrix} 7 & 0 & 3 \\ 5 & 1 & 2 \\ 2 & 0 & 1 \end{pmatrix} \begin{pmatrix} x_1 \\ x_2 \\ x_2 \end{pmatrix} = \begin{pmatrix} 2 \\ 3 \\ -1 \end{pmatrix}$

c) $\begin{pmatrix} 1 & 1 & 0 \\ 1 & 0 & 1 \\ 1 & 1 & 1 \end{pmatrix} \begin{pmatrix} x_1 \\ x_2 \\ x_3 \end{pmatrix} = \begin{pmatrix} 3 \\ 1 \\ -1 \end{pmatrix}$

d) $\begin{pmatrix} 3 & 3 & 0 & 1 \\ 1 & -1 & 1 & 0 \\ 1 & -2 & 1 & 0 \\ 2 & 3 & 0 & 1 \end{pmatrix} \begin{pmatrix} x_1 \\ x_2 \\ x_3 \\ x_4 \end{pmatrix} = \begin{pmatrix} 1 \\ 2 \\ -2 \\ -1 \end{pmatrix}$.

23. Use row reduction to show that the following matrices are singular

a) $\begin{pmatrix} 3 & 6 \\ 6 & 12 \end{pmatrix}$ **b)** $\begin{pmatrix} 1 & 2 & -4 \\ -1 & 1 & 1 \\ 1 & 5 & -7 \end{pmatrix}$ **c)** $\begin{pmatrix} 1 & 1 & 0 & 1 \\ 1 & 1 & 1 & 0 \\ 1 & 2 & 1 & 3 \\ 2 & 1 & 0 & -1 \end{pmatrix}$

d) $\begin{pmatrix} 1 & 1 & 1 & 1 & -1 \\ 2 & 1 & -3 & 1 & 10 \\ 1 & 1 & 0 & 1 & 6 \\ 1 & -1 & 1 & -1 & -5 \\ 4 & 1 & 0 & 1 & 9 \end{pmatrix}$.

24. Let $Q = \frac{1}{11}\begin{pmatrix} 9 & -2 & -6 \\ -2 & 9 & -6 \\ 6 & 6 & 7 \end{pmatrix}$ represent the change of coordinates from (x, y, z) to (x', y', z'). Convert the following from (x', y', z') coordinates to (x, y, z) coordinates.

a) $(4,4,1)$ **b)** $(6,2,1)$ **c)** $(5,5,4)$.

25. Let

(i)
$$\begin{aligned} x + 2y + z &= u \\ -x + y - z &= v \end{aligned}$$

and

(ii)
$$\begin{aligned} 2u + v &= f \\ u - 2v &= g \\ -u + v &= h \end{aligned} .$$

a) Write systems (i) & (ii) in matrix form

$$A\begin{pmatrix} x \\ y \\ z \end{pmatrix} = \begin{pmatrix} u \\ v \end{pmatrix} \quad \& \quad B\begin{pmatrix} u \\ v \end{pmatrix} = \begin{pmatrix} f \\ g \\ h \end{pmatrix} .$$

h) Use the results of **a)** to compute the matrix M such that

$$M\begin{pmatrix} x \\ y \\ z \end{pmatrix} = \begin{pmatrix} f \\ g \\ h \end{pmatrix}.$$

26. Compute the following

a) $\begin{pmatrix} 2 & -1 \\ 1 & 3 \end{pmatrix}\begin{pmatrix} 1 & -1 & 5 & -1 \\ 1 & 4 & 2 & 3 \end{pmatrix} + \begin{pmatrix} -2 & 3 \\ 1 & -3 \end{pmatrix}\begin{pmatrix} 1 & -1 & 5 & -1 \\ 1 & 4 & 2 & 3 \end{pmatrix}$

b) $2\begin{pmatrix} 2 \\ -1 \\ 3 \end{pmatrix}\begin{pmatrix} 3 \\ 5 \\ 1 \end{pmatrix}^T + \begin{pmatrix} 7 \\ -2 \\ 3 \end{pmatrix}\begin{pmatrix} -1 \\ 2 \\ 1 \end{pmatrix}^T$

c) $2\begin{pmatrix} 3 \\ 5 \\ 1 \end{pmatrix}^T\begin{pmatrix} 2 \\ -1 \\ 3 \end{pmatrix} + \begin{pmatrix} -1 \\ 2 \\ 1 \end{pmatrix}^T\begin{pmatrix} 7 \\ -2 \\ 3 \end{pmatrix}.$

27. Compute

$$\begin{vmatrix} 1 & -1 & 1 & 1 \\ 2 & 1 & 4 & -1 \\ 4 & -1 & 2 & 3 \\ 3 & 1 & -1 & 2 \end{vmatrix}.$$

28. Compute the $(3,1)$ cofactor, A_{31}, of

$$\begin{vmatrix} 1 & 2 & -1 & 3 \\ -1 & 3 & 2 & 1 \\ 2 & -1 & 3 & 7 \\ 1 & 2 & 1 & -1 \end{vmatrix}.$$

29. Suppose that

$$A^{-1} = \begin{pmatrix} 13 & -8 & 3 \\ -9 & 7 & -3 \\ 2 & -2 & 1 \end{pmatrix}.$$

Find A.

30. Assume that

$$Q = \frac{1}{7}\begin{pmatrix} 6 & q_{12} & q_{13} \\ 3 & 6 & q_{23} \\ -2 & q_{32} & q_{33} \end{pmatrix}$$

is an orthogonal matrix with determinant one and $q_{32} > 0$. Find q_{12}, q_{13}, q_{23}, q_{32} and q_{33}.

31. Suppose that $A = \begin{pmatrix} -5 & 1 & -2 \\ 17 & -3 & 7 \\ 55 & -10 & 22 \end{pmatrix}$ and $XA = I_3$. Find X.

32.

$$S = \begin{pmatrix} 3 & -1 & 2 \\ -1 & 3 & -2 \\ 2 & -2 & 6 \end{pmatrix}$$

then $\lambda_3 = \lambda_1 = 2$ is a twice repeated eigenvalue.

a) Find the 3rd eigenvalue λ_2 without computing a characteristic equation. First find an ***eigenvector*** $\mathbf{v}_2$ corresponding to λ_2 and then find λ_2 by using matrix multiplication to directly verify that $\mathbf{v}_2$ is indeed an eigenvector.

b)

$$\mathbf{v}_3 = \begin{pmatrix} -1 \\ 1 \\ 1 \end{pmatrix}$$

is one eigenvector corresponding to $\lambda_3 = 2$. Find another ***eigenvector*** corresponding to $\lambda_1 = 2$ such that $\{\mathbf{v}_1, \mathbf{v}_2, \mathbf{v}_3\}$ is a right handed set of eigenvectors.

10 The Chain Rule in Matrix Form

Outline of Key Ideas

Chain rule for $y = f(u(x))$.

Chain rule for $z = f(x, y)$:

$\mathbf{R} = \langle x, y \rangle = \mathbf{R}(t)$ is a point on a curve

$\mathbf{R} = \langle x, y \rangle = \mathbf{R}(u, v)$ gives a mapping from $\mathcal{R}^2$ to $\mathcal{R}^2$.

Chain rule for $g = f(x, y, z)$:

$\mathbf{R} = \langle x, y, z \rangle = \mathbf{R}(t)$ is a point on a curve

$\mathbf{R} = \langle x, y, z \rangle = \mathbf{R}(u, v)$ is a point on a surface

$\mathbf{R} = \langle x, y, z \rangle = \mathbf{R}(u, v, w)$ gives a mapping from $\mathcal{R}^3$ to $\mathcal{R}^3$.

General chain rule for $g = f(x_1, \ldots, x_m)$ where $\mathbf{R} = \langle x_1, \ldots, x_m \rangle = \mathbf{R}(u_1, \ldots, u_n)$.

[illegible]

Outline of Key Ideas

[illegible]

The Chain Rule in Matrix Form

Matrices provide a unified way to view and interpret the chain rule. We will quickly treat the one variable case and then proceed to the several variable case.

10.1 Chain Rule for $y = f(u(x))$

If $f = f(u)$ and $u = u(x)$ then $y = y(x) = f(u(x))$ is a function of x and its derivative $\dfrac{dy}{dx}$ is given by the chain rule

$$\frac{dy}{dx} = \frac{df}{du}\frac{du}{dx} = f'(u)\frac{du}{dx} = f'(u(x))\frac{du}{dx} \tag{10.1}$$

which says that $\dfrac{dy}{dx}$ equals the product of $f'(u)$ at $u = u(x)$ and $\dfrac{du}{dx}$. For example if $y = f(u) = \sin u$ and $u = \ln x$ then $f'(u) = \cos u = \cos(\ln x)$ and $\dfrac{du}{dx} = \dfrac{1}{x}$ so $\dfrac{dy}{dx}$ is given by

$$\frac{dy}{dx} = \frac{1}{x}\cos(\ln x) \quad .$$

10.2 Chain Rule for $z = f(x, y)$

If $f = f(x, y)$ is a function of two variables then $\mathbf{R} = \langle x, y\rangle$ could be a point on a curve $\mathbf{R} = \mathbf{R}(t)$ or could correspond to a mapping from $\mathfrak{R}^2$ to $\mathfrak{R}^2$ given by $\mathbf{R} = \mathbf{R}(u, v)$.

a) If $\mathbf{R} = \mathbf{R}(t)$ is a point on a curve then $z = z(t) = f(x(t), y(t))$ is a function of t and by the chain rule

$$\frac{dz}{dt} = \frac{\partial f}{\partial x}\frac{dx}{dt} + \frac{\partial f}{\partial y}\frac{dy}{dt} = f_x\frac{dx}{dt} + f_y\frac{dy}{dt} = \nabla f \bullet \frac{d\mathbf{R}}{dt} \tag{10.2}$$

in vector form where

$$\nabla f = \left\langle \frac{\partial f}{\partial x}, \frac{\partial f}{\partial y} \right\rangle = \langle f_x, f_y\rangle$$

is the ***gradient*** of f and

$$\frac{d\mathbf{R}}{dt} = \left\langle \frac{dx}{dt}, \frac{dy}{dt} \right\rangle$$

is the ***velocity vector*** to the curve $\mathbf{R} = \mathbf{R}(t)$. In matrix form $\dfrac{dz}{dt}$ can be expressed by

$$(10.3) \qquad \frac{dz}{dt} = \begin{pmatrix} \dfrac{\partial f}{\partial x} & \dfrac{\partial f}{\partial y} \end{pmatrix} \begin{pmatrix} \dfrac{dx}{dt} \\ \dfrac{dy}{dt} \end{pmatrix} = \begin{pmatrix} f_x & f_y \end{pmatrix} \begin{pmatrix} \dfrac{dx}{dt} \\ \dfrac{dy}{dt} \end{pmatrix}$$

where ∇f is represented by the 1×2 row vector

$$\begin{pmatrix} \dfrac{\partial f}{\partial x} & \dfrac{\partial f}{\partial y} \end{pmatrix} = \begin{pmatrix} f_x & f_y \end{pmatrix}$$

and $\dfrac{d\mathbf{R}}{dt}$ is represented by the 2×1 column vector $\left[\dfrac{dx}{dt}, \dfrac{dy}{dt}\right]^T$.

b) If $\mathbf{R} = \mathbf{R}(u, v)$ is a mapping from $\mathcal{R}^2$ to $\mathcal{R}^2$ then $z = z(u,v) = f(x(u,v), y(u,v))$ is a function of u and v, and $\dfrac{\partial z}{\partial u}$ and $\dfrac{\partial z}{\partial v}$ are given by the following:

$$(10.4) \qquad \begin{aligned} \frac{\partial z}{\partial u} &= \frac{\partial f}{\partial x}\frac{\partial x}{\partial u} + \frac{\partial f}{\partial y}\frac{\partial y}{\partial u} = f_x \frac{\partial x}{\partial u} + f_y \frac{\partial y}{\partial u} = \nabla f \bullet \frac{\partial \mathbf{R}}{\partial u} \\ \frac{\partial z}{\partial v} &= \frac{\partial f}{\partial x}\frac{\partial x}{\partial v} + \frac{\partial f}{\partial y}\frac{\partial y}{\partial v} = f_x \frac{\partial x}{\partial v} + f_y \frac{\partial y}{\partial v} = \nabla f \bullet \frac{\partial \mathbf{R}}{\partial v} \end{aligned}.$$

In matrix form $\dfrac{\partial z}{\partial u}$ and $\dfrac{\partial z}{\partial v}$ can be expressed by the matrix equations

$$\frac{\partial z}{\partial u} = \begin{pmatrix} \dfrac{\partial f}{\partial x} & \dfrac{\partial f}{\partial y} \end{pmatrix} \begin{pmatrix} \dfrac{\partial x}{\partial u} \\ \dfrac{\partial y}{\partial u} \end{pmatrix} = \begin{pmatrix} f_x & f_y \end{pmatrix} \begin{pmatrix} \dfrac{\partial x}{\partial u} \\ \dfrac{\partial y}{\partial u} \end{pmatrix}$$

$$\frac{\partial z}{\partial v} = \begin{pmatrix} \dfrac{\partial f}{\partial x} & \dfrac{\partial f}{\partial y} \end{pmatrix} \begin{pmatrix} \dfrac{\partial x}{\partial v} \\ \dfrac{\partial y}{\partial v} \end{pmatrix} = \begin{pmatrix} f_x & f_y \end{pmatrix} \begin{pmatrix} \dfrac{\partial x}{\partial v} \\ \dfrac{\partial y}{\partial v} \end{pmatrix}$$

which can be combined into one matrix equation

$$(10.5)\qquad \begin{pmatrix} \frac{\partial z}{\partial u} & \frac{\partial z}{\partial v} \end{pmatrix} = \begin{pmatrix} \frac{\partial f}{\partial x} & \frac{\partial f}{\partial y} \end{pmatrix} \begin{pmatrix} \frac{\partial x}{\partial u} & \frac{\partial x}{\partial v} \\ \frac{\partial y}{\partial u} & \frac{\partial y}{\partial v} \end{pmatrix} =$$

$$\begin{pmatrix} f_x & f_y \end{pmatrix} \begin{pmatrix} \frac{\partial x}{\partial u} & \frac{\partial x}{\partial v} \\ \frac{\partial y}{\partial u} & \frac{\partial y}{\partial v} \end{pmatrix}$$

which expresses the 1×2 row vector $\left[\frac{\partial z}{\partial u}, \frac{\partial z}{\partial v}\right]$ as a product of the 1×2 row vector $[f_x, f_y]$ times the 2×2 matrix

$$\begin{pmatrix} \frac{\partial x}{\partial u} & \frac{\partial x}{\partial v} \\ \frac{\partial y}{\partial u} & \frac{\partial y}{\partial v} \end{pmatrix}$$

whose columns are the coordinate columns of $\frac{\partial \mathbf{R}}{\partial u}$ and $\frac{\partial \mathbf{R}}{\partial v}$. As we shall see Equations (10.3) and (10.5) show the general form of the chain rule for functions of several variables.

10.3 Chain Rule for $g = f(x, y, z)$

If $f = f(x, y, z)$ then $\mathbf{R} = \langle x, y, z\rangle$ could be a point on a curve $\mathbf{R} = \mathbf{R}(t)$, could be a point on a surface $\mathbf{R} = \mathbf{R}(u, v)$, or could correspond to a mapping from $\mathcal{R}^3$ to $\mathcal{R}^3$ given by $\mathbf{R} = \mathbf{R}(u, v, w)$.

a) If $\mathbf{R} = \mathbf{R}(t)$ is a point on a curve then $z = z(t) = f(x(t), y(t), z(t))$ is a function of t and by the chain rule

$$(10.6)\qquad \frac{dz}{dt} = \frac{\partial f}{\partial x}\frac{dx}{dt} + \frac{\partial f}{\partial y}\frac{dy}{dt} + \frac{\partial f}{\partial z}\frac{dz}{dt} = f_x\frac{dx}{dt} + f_y\frac{dy}{dt} + f_z\frac{dz}{dt} = \nabla f \bullet \frac{d\mathbf{R}}{dt}$$

where

$$\nabla f = \left\langle \frac{\partial f}{\partial x}, \frac{\partial f}{\partial y}, \frac{\partial f}{\partial z} \right\rangle = \langle f_x, f_y, f_z \rangle$$

is the ***gradient*** of f and $\frac{d\mathbf{R}}{dt} = \left\langle \frac{dx}{dt}, \frac{dy}{dt}, \frac{dz}{dt} \right\rangle$ is the ***velocity vector*** to the curve $\mathbf{R} = \mathbf{R}(t)$.

The matrix form of the chain rule for $\frac{dg}{dt}$ can be expressed by

$$(10.7) \qquad \frac{dg}{dt} = \begin{pmatrix} \frac{\partial f}{\partial x} & \frac{\partial f}{\partial y} & \frac{\partial f}{\partial z} \end{pmatrix} \begin{pmatrix} \frac{dx}{dt} \\ \frac{dy}{dt} \\ \frac{dz}{dt} \end{pmatrix}$$

$$= \begin{pmatrix} f_x & f_y & f_z \end{pmatrix} \begin{pmatrix} \frac{dx}{dt} \\ \frac{dy}{dt} \\ \frac{dz}{dt} \end{pmatrix}$$

where ∇f is represented by the 1×3 row vector

$$\begin{pmatrix} \frac{\partial f}{\partial x} & \frac{\partial f}{\partial y} & \frac{\partial f}{\partial z} \end{pmatrix} = \begin{pmatrix} f_x & f_y & f_z \end{pmatrix}$$

and $\frac{d\mathbf{R}}{dt}$ is represented by its coordinate column

$$\begin{pmatrix} \frac{dx}{dt} & \frac{dy}{dt} & \frac{dz}{dt} \end{pmatrix}^T .$$

b) If $\mathbf{R} = \mathbf{R}(u, v)$ is a point on a surface then

$$g = g(u, v) = f(x(u, v), y(u, v), z(u, v))$$

is a function of (u, v) and the matrix form of the chain rule for $\left[\frac{\partial g}{\partial u}, \frac{\partial g}{\partial v}\right]$ is given by

(10.8)
$$\begin{pmatrix} \frac{\partial g}{\partial u} & \frac{\partial g}{\partial v} \end{pmatrix} = \begin{pmatrix} \frac{\partial f}{\partial x} & \frac{\partial f}{\partial y} & \frac{\partial f}{\partial z} \end{pmatrix} \begin{pmatrix} \frac{\partial x}{\partial u} & \frac{\partial x}{\partial v} \\ \frac{\partial y}{\partial u} & \frac{\partial y}{\partial v} \\ \frac{\partial z}{\partial u} & \frac{\partial z}{\partial v} \end{pmatrix}$$

$$= \begin{pmatrix} f_x & f_y & f_z \end{pmatrix} \begin{pmatrix} \frac{\partial x}{\partial u} & \frac{\partial x}{\partial v} \\ \frac{\partial y}{\partial u} & \frac{\partial y}{\partial v} \\ \frac{\partial z}{\partial u} & \frac{\partial z}{\partial v} \end{pmatrix} .$$

c) If $\mathbf{R} = \mathbf{R}(u,v,w)$ is a mapping from $\mathcal{R}^3$ to $\mathcal{R}^3$ then $g = g(u,v,w) = f(x(u,v,w), y(u,v,w), z(u,v,w))$ is a function of (u,v,w) and the matrix form of the chain rule for $\left[\frac{\partial g}{\partial u}, \frac{\partial g}{\partial v}, \frac{\partial g}{\partial w}\right]$ is given by

(10.9)
$$\begin{pmatrix} \frac{\partial g}{\partial u} & \frac{\partial g}{\partial v} & \frac{\partial g}{\partial w} \end{pmatrix}$$

$$= \begin{pmatrix} \frac{\partial f}{\partial x} & \frac{\partial f}{\partial y} & \frac{\partial f}{\partial z} \end{pmatrix} \begin{pmatrix} \frac{\partial x}{\partial u} & \frac{\partial x}{\partial v} & \frac{\partial x}{\partial w} \\ \frac{\partial y}{\partial u} & \frac{\partial y}{\partial v} & \frac{\partial y}{\partial w} \\ \frac{\partial z}{\partial u} & \frac{\partial z}{\partial v} & \frac{\partial z}{\partial w} \end{pmatrix}$$

$$= \begin{pmatrix} f_x & f_y & f_z \end{pmatrix} \begin{pmatrix} \frac{\partial x}{\partial u} & \frac{\partial x}{\partial v} & \frac{\partial x}{\partial w} \\ \frac{\partial y}{\partial u} & \frac{\partial y}{\partial v} & \frac{\partial y}{\partial w} \\ \frac{\partial z}{\partial u} & \frac{\partial z}{\partial v} & \frac{\partial z}{\partial w} \end{pmatrix} .$$

10.4 General Chain Rule for $g = f(x_1, \ldots, x_m)$ Where $\mathbf{R} = \langle x_1, \ldots, x_m \rangle = \mathbf{R}(u_1, \ldots, u_n)$

From the preceding we are ready to present the general matrix form of the chain rule.

If $f = f(x_1, \ldots, x_m)$ and if each variable $x_i = x_i(u_1, \ldots, u_n)$, $i = 1, \ldots, m$ is a function of $(u_1, \ldots, u_n)$ then $\mathbf{R} = \mathbf{R}(u_1, \ldots, u_n)$ and

$$g = g(u_1, \ldots, u_n) = f(x_1(u_1, \ldots, u_n), \ldots, x_m(u_1, \ldots, u_n))$$

is a function of $(u_1, \ldots, u_n)$. Then the matrix form of the chain rule for

$$\left[\frac{\partial g}{\partial u_1}, \ldots, \frac{\partial g}{\partial u_n}\right]$$

is given by

$$(10.10) \qquad \begin{pmatrix} \dfrac{\partial g}{\partial u_1} & \cdots & \dfrac{\partial g}{\partial u_n} \end{pmatrix}$$

$$= \begin{pmatrix} \dfrac{\partial f}{\partial x_1} & \cdots & \dfrac{\partial f}{\partial x_m} \end{pmatrix} \begin{pmatrix} \dfrac{\partial x_1}{\partial u_1} & \cdots & \dfrac{\partial x_1}{\partial u_n} \\ \vdots & & \vdots \\ \dfrac{\partial x_m}{\partial u_1} & \cdots & \dfrac{\partial x_m}{\partial u_n} \end{pmatrix}$$

where $\left[\dfrac{\partial g}{\partial u_1}, \ldots, \dfrac{\partial g}{\partial u_n}\right]$ is a $1 \times n$ row vector, the ***gradient*** of $g = g(u_1, \ldots, u_n)$ with respect to $(u_1, \ldots, u_n)$,

$$\nabla f = \left[\frac{\partial f}{\partial x_1}, \ldots, \frac{\partial f}{\partial x_m}\right]$$

is a $1 \times m$ row vector, the ***gradient*** of $f = f(x_1, \ldots, x_m)$ with respect to $(x_1, \ldots, x_m)$, and $\left[\dfrac{\partial x_i}{\partial u_j}\right]_{m \times n}$ is an $m \times n$ matrix whose j^{th} column is the coordinate column of $\dfrac{\partial \mathbf{R}}{\partial u_j}$.

The jth component of $\left[\dfrac{\partial g}{\partial u_1}, \ldots, \dfrac{\partial g}{\partial u_n}\right]$ is $\dfrac{\partial g}{\partial u_j}$ and is given by

$$(10.11)\qquad \frac{\partial g}{\partial u_j} = \begin{pmatrix} \frac{\partial f}{\partial x_1} & \cdots & \frac{\partial f}{\partial x_m} \end{pmatrix} \begin{pmatrix} \frac{\partial x_1}{\partial u_j} \\ \vdots \\ \frac{\partial x_m}{\partial u_j} \end{pmatrix} = \nabla f \bullet \frac{\partial \mathbf{R}}{\partial u_j} \quad .$$

The following examples will illustrate these ideas.

Example 10.1

Let $f(x, y) = x^2 - xy + y^2$ and let $\mathbf{R} = \langle x, y\rangle$ be the curve $\mathbf{R} = \langle t^3, t^5\rangle$. Then $z = t^6 - t^8 + t^{10}$ is a function of t and $\frac{dz}{dt} = 6t^5 - 8t^7 + 10t^9$. Verify the chain rule

$$\frac{dz}{dt} = \begin{pmatrix} f_x & f_y \end{pmatrix} \begin{pmatrix} \frac{dx}{dt} \\ \frac{dy}{dt} \end{pmatrix} \quad .$$

Solution.

$$\begin{pmatrix} f_x & f_y \end{pmatrix} = \begin{pmatrix} 2x - y, & -x + 2y \end{pmatrix} = \begin{pmatrix} 2t^3 - t^5, & -t^3 + 2t^5 \end{pmatrix}$$

and

$$\begin{pmatrix} \frac{dx}{dt} \\ \frac{dy}{dt} \end{pmatrix} = \begin{pmatrix} 3t^2 \\ 5t^4 \end{pmatrix} \quad .$$

Thus,

$$\begin{aligned} \begin{pmatrix} f_x & f_y \end{pmatrix} \begin{pmatrix} \frac{dx}{dt} \\ \frac{dy}{dt} \end{pmatrix} &= \begin{pmatrix} 2t^3 - t^5, & -t^3 + 2t^5 \end{pmatrix} \begin{pmatrix} 3t^2 \\ 5t^4 \end{pmatrix} \\ &= (2t^3 - t^5)(3t^2) + (-t^3 + 2t^5)(5t^4) \\ &= 6t^5 - 8t^7 + 10t^9 = \frac{dz}{dt} \quad . \blacksquare \end{aligned}$$

Example 10.2

Let $f(x,y) = x^3 - xy^2$ and let a mapping from $\mathcal{R}^2$ to $\mathcal{R}^2$ be given by $\mathbf{R} = \langle x, y \rangle = \langle uv, v \rangle$. Then

$$z = f(uv, v) = (uv)^3 - (uv)v^2 = u^3v^3 - uv^3$$

is a function of (u, v), and $\left[\frac{\partial z}{\partial u}, \frac{\partial z}{\partial v}\right]$ is given by

$$\begin{pmatrix} \frac{\partial z}{\partial u} & \frac{\partial z}{\partial v} \end{pmatrix} = \begin{pmatrix} 3u^2v^3 - v^3, & 3u^3v^2 - 3uv^2 \end{pmatrix} .$$

Verify the chain rule

$$\begin{pmatrix} \frac{\partial z}{\partial u} & \frac{\partial z}{\partial v} \end{pmatrix} = \begin{pmatrix} f_x & f_y \end{pmatrix} \begin{pmatrix} \frac{\partial x}{\partial u} & \frac{\partial x}{\partial v} \\ \frac{\partial y}{\partial u} & \frac{\partial y}{\partial v} \end{pmatrix} .$$

Solution.

$$\begin{pmatrix} f_x & f_y \end{pmatrix} = \begin{pmatrix} 3x^2 - y^2, & -2xy \end{pmatrix} = \begin{pmatrix} 3u^2v^2 - v^2, & -2uv^2 \end{pmatrix}$$

and

$$\begin{pmatrix} \frac{\partial x}{\partial u} & \frac{\partial x}{\partial v} \\ \frac{\partial y}{\partial u} & \frac{\partial y}{\partial v} \end{pmatrix} = \begin{pmatrix} v & u \\ 0 & 1 \end{pmatrix} .$$

Thus,

$$\begin{pmatrix} f_x & f_y \end{pmatrix} \begin{pmatrix} \frac{\partial x}{\partial u} & \frac{\partial x}{\partial v} \\ \frac{\partial y}{\partial u} & \frac{\partial y}{\partial v} \end{pmatrix} = \begin{pmatrix} 3u^2v^2 - v^2, & -2uv^2 \end{pmatrix} \begin{pmatrix} v & u \\ 0 & 1 \end{pmatrix}$$

$$= \begin{pmatrix} (3u^2v^2 - v^2)v, & (3u^2v^2 - v^2)u - 2uv^2 \end{pmatrix}$$

$$= \begin{pmatrix} 3u^2v^3 - v^3, & 3u^3v^2 - 3uv^2 \end{pmatrix}$$

$$= \begin{pmatrix} \frac{\partial z}{\partial u} & \frac{\partial z}{\partial v} \end{pmatrix} . \blacksquare$$

Example 10.3

Let $f(x,y,z) = 2x^2y - xz$ and let $\mathbf{R} = \langle x,y,z\rangle$ be the curve $\mathbf{R} = \langle t,t^2,t^3\rangle$. Then

$$g = f(t,t^2,t^3) = 2t^4 - t^4 = t^4$$

is a function of t. Verify the chain rule

$$\frac{dg}{dt} = \begin{pmatrix} \frac{\partial f}{\partial x} & \frac{\partial f}{\partial y} & \frac{\partial f}{\partial z} \end{pmatrix} \begin{pmatrix} \frac{dx}{dt} \\ \frac{dy}{dt} \\ \frac{dz}{dt} \end{pmatrix} .$$

Solution.

$$\begin{pmatrix} f_x & f_y & f_z \end{pmatrix} = \begin{pmatrix} 4xy - z, & 2x^2, & -x \end{pmatrix} = \begin{pmatrix} 4t^3 - t^3, & 2t^2, & -t \end{pmatrix}$$

$$= \begin{pmatrix} 3t^3, & 2t^2, & -t \end{pmatrix}$$

and

$$\begin{pmatrix} \frac{dx}{dt} \\ \frac{dy}{dt} \\ \frac{dz}{dt} \end{pmatrix} = \begin{pmatrix} 1 \\ 2t \\ 3t^2 \end{pmatrix} .$$

Thus,

$$\begin{pmatrix} 3t^3 & 2t^2 & -t \end{pmatrix} \begin{pmatrix} 1 \\ 2t \\ 3t^2 \end{pmatrix} = 3t^3 + 4t^3 - 3t^3 = 4t^3 = \frac{dg}{dt} . \blacksquare$$

Example 10.4

Let $f(x, y, z) = 2x^2y - xz$ and let $\mathbf{R} = \langle x, y, z \rangle$ be the surface $\mathbf{R} = \langle \cos u, \sin u, v \rangle$. Then

$$g = f(\cos u, \sin u, v) = 2 \sin u \cos^2 u - v \cos u$$

is a function of u and v. Verify the chain rule

$$\begin{pmatrix} \dfrac{\partial g}{\partial u} & \dfrac{\partial g}{\partial v} \end{pmatrix} = \begin{pmatrix} \dfrac{\partial f}{\partial x} & \dfrac{\partial f}{\partial y} & \dfrac{\partial f}{\partial z} \end{pmatrix} \begin{pmatrix} \dfrac{\partial x}{\partial u} & \dfrac{\partial x}{\partial v} \\ \dfrac{\partial y}{\partial u} & \dfrac{\partial y}{\partial v} \\ \dfrac{\partial z}{\partial u} & \dfrac{\partial z}{\partial v} \end{pmatrix} .$$

Solution.

$$\begin{pmatrix} f_x & f_y & f_z \end{pmatrix}$$

$$= \begin{pmatrix} 4xy - z, & 2x^2, & -x \end{pmatrix} = \begin{pmatrix} 4 \cos u \sin u - v, & 2 \cos^2 u, & -\cos u \end{pmatrix}$$

and

$$\begin{pmatrix} \dfrac{\partial x}{\partial u} & \dfrac{\partial x}{\partial v} \\ \dfrac{\partial y}{\partial u} & \dfrac{\partial y}{\partial v} \\ \dfrac{\partial z}{\partial u} & \dfrac{\partial z}{\partial v} \end{pmatrix} = \begin{pmatrix} -\sin u & 0 \\ \cos u & 0 \\ 0 & 1 \end{pmatrix} .$$

Thus,

$$\begin{pmatrix} 4 \cos u \sin u - v, & 2 \cos^2 u, & -\cos u \end{pmatrix} \begin{pmatrix} -\sin u & 0 \\ \cos u & 0 \\ 0 & 1 \end{pmatrix}$$

$$\begin{pmatrix} -4 \cos u \sin^2 u + v \sin u + 2 \cos^3 u, & -\cos u \end{pmatrix} = \begin{pmatrix} g_u & g_v \end{pmatrix} . \ \blacksquare$$

Example 10.5

Let $f(x,y,z) = xy - z^2$ and let $\mathbf{R} = \langle x,y,z\rangle$ correspond to the mapping from $\mathcal{R}^3$ to $\mathcal{R}^3$ given by $\mathbf{R} = \langle u^2\cosh w, v^2\cosh w, uv\sinh w\rangle$. Then

$$\begin{aligned} g &= g(u,v,w) = f(u^2\cosh w, v^2\cosh w, uv\sinh w) \\ &= u^2v^2\cosh^2 w - u^2v^2\sinh^2 w = u^2v^2 \end{aligned}$$

is a function of (u,v,w) and $\langle g_u, g_v, g_w\rangle$ is given by $\langle 2uv^2, 2u^2v, 0\rangle$. To verify the chain rule we have

$$\langle f_x, f_y, f_z\rangle = \langle y, x, -2z\rangle = \langle v^2\cosh w, u^2\cosh w, -2uv\sinh w\rangle$$

and

$$\begin{pmatrix} \dfrac{\partial x}{\partial u} & \dfrac{\partial x}{\partial v} & \dfrac{\partial x}{\partial w} \\ \dfrac{\partial y}{\partial u} & \dfrac{\partial y}{\partial v} & \dfrac{\partial y}{\partial w} \\ \dfrac{\partial z}{\partial u} & \dfrac{\partial z}{\partial v} & \dfrac{\partial z}{\partial w} \end{pmatrix} = \begin{pmatrix} 2u\cosh w & 0 & u^2\sinh w \\ 0 & 2v\cosh w & v^2\sinh w \\ v\sinh w & u\sinh w & uv\cosh w \end{pmatrix}.$$

Then

$$\begin{pmatrix} v^2\cosh w, & u^2\cosh w, & -2uv\sinh w \end{pmatrix} \begin{pmatrix} 2u\cosh w & 0 & u^2\sinh w \\ 0 & 2v\cosh w & v^2\sinh w \\ v\sinh w & u\sinh w & uv\cosh w \end{pmatrix}$$

$$= \left[2uv^2(\cosh^2 w - \sinh^2 w), 2u^2v(\cosh^2 w - \sinh^2 w), (2u^2v^2 - 2u^2v^2)\sinh w\cosh w\right]$$

$$= \left[2uv^2, 2u^2v, 0\right] = \left[g_u, g_v, g_w\right]. \quad \blacksquare$$

Example 10.6

Let $f(u) = \cos u$ and let $u = x^2y$. Then $g = g(x,y) = f(x^2y) = \cos(x^2y)$ and $\nabla g = \langle g_x, g_y\rangle$ is given by $\langle -2xy\sin(x^2y), -x^2\sin(x^2y)\rangle = -\sin(x^2y)\langle 2xy, x^2\rangle$. Verify the chain rule for $\langle g_x, g_y\rangle$ in matrix form

$$\begin{pmatrix} \dfrac{\partial g}{\partial x} & \dfrac{\partial g}{\partial y} \end{pmatrix} = f'(u)\begin{pmatrix} \dfrac{\partial u}{\partial x} & \dfrac{\partial u}{\partial y} \end{pmatrix}.$$

Solution.

$$f'(u) = -\sin u = -\sin(x^2 y)$$

and

$$\begin{pmatrix} \frac{\partial u}{\partial x} & \frac{\partial u}{\partial y} \end{pmatrix} = \begin{pmatrix} 2xy & x^2 \end{pmatrix} .$$

Thus,

$$f'(u)\left\langle \frac{\partial u}{\partial x}, \frac{\partial u}{\partial y} \right\rangle = -\sin(x^2 y)\langle 2xy, x^2 \rangle = \langle g_x, g_y \rangle \quad . \blacksquare$$

Example 10.7

Let $f(x,y) = 3x - 5y$ and let $\mathbf{R} = \langle x, y \rangle$ be given by $\mathbf{R} = \langle 2u - 2v + w, u + 2v + 2w \rangle$. Then $g = g(u,v,w) = f(2u - 2v + w, u + 2v + 2w) = u - 16v - 7w$ is a function of (u,v,w) and $\langle g_u, g_v, g_w \rangle = \langle 1, -16, -7 \rangle$. Verify the chain rule for $\langle g_u, g_v, g_w \rangle$ in matrix form

$$\begin{pmatrix} \frac{\partial g}{\partial u} & \frac{\partial g}{\partial v} & \frac{\partial g}{\partial w} \end{pmatrix} = \begin{pmatrix} \frac{\partial f}{\partial x} & \frac{\partial f}{\partial y} \end{pmatrix} \begin{pmatrix} \frac{\partial x}{\partial u} & \frac{\partial x}{\partial v} & \frac{\partial x}{\partial w} \\ \frac{\partial y}{\partial u} & \frac{\partial y}{\partial v} & \frac{\partial y}{\partial w} \end{pmatrix} .$$

Solution.

$$\nabla f = \langle f_x, f_y \rangle = \langle 3, -5 \rangle$$

and

$$\begin{pmatrix} \frac{\partial x}{\partial u} & \frac{\partial x}{\partial v} & \frac{\partial x}{\partial w} \\ \frac{\partial y}{\partial u} & \frac{\partial y}{\partial v} & \frac{\partial y}{\partial w} \end{pmatrix} = \begin{pmatrix} 2 & -2 & 1 \\ 1 & 2 & 2 \end{pmatrix} .$$

Thus,

$$\begin{pmatrix} 3 & -5 \end{pmatrix} \begin{pmatrix} 2 & -2 & 1 \\ 1 & 2 & 2 \end{pmatrix} = \begin{pmatrix} 1 & -16 & -7 \end{pmatrix} = \begin{pmatrix} g_u & g_v & g_w \end{pmatrix} \quad . \blacksquare$$

10.5 Exercises

1. Find $f'(u)$ for each of the following if

a) $f(u) - e^u$ and $u = x^2y^3$

b) $f(u) = \arctan u$ and $u = \sinh(x^2 + y^3 + z)$

c) $f(u) = \ln u$ and $u = x^2 + y^4$.

2. For each of the following find the $m \times n$ matrix $\left[\dfrac{\partial x_i}{\partial u_j}\right]$ if $\mathbf{R} = \langle x_1, \ldots, x_m \rangle = \mathbf{R}(u_1, \ldots, u_n)$ is given by

a) $\mathbf{R} = \langle u_1 \cos u_2, u_1 \cos u_2, u_3 \rangle$

b) $\mathbf{R} = \langle u_1 \sin u_2 \cos u_3, u_1 \sin u_2 \sin u_3, u_1 \cos u_2 \rangle$

c) $\mathbf{R} = \langle a \cos u_1 \cos u_2, a \sin u_1 \cos u_2, a \sin u_2 \rangle$

d) $\mathbf{R} = \langle u_1^2 - u_2^2, 2u_1u_2 \rangle$

e) $\mathbf{R} = \langle 2u_1 + 3u_2, u_1 + u_2 \rangle$

f) $\mathbf{R} = \langle 2u_1 + u_2 - u_3, -u_1 + 2u_2 + 3u_3 \rangle$

g) $\mathbf{R} = \langle -u_1 + 2u_2 + 3u_3, u_1 + u_2 - u_3, 2u_1 - 3u_2 + 2u_3 \rangle$.

3. If $f(x_1, \ldots, x_m)$ and $\mathbf{R} = \langle x_1, \ldots, x_m \rangle = \mathbf{R}(u_1, \ldots, u_n)$ express

$$\nabla f = \left\langle \frac{\partial f}{\partial x_1}, \ldots, \frac{\partial f}{\partial x_m} \right\rangle$$

in terms of $(u_1, \ldots, u_n)$ if

a) $f(x_1, x_2) = x_1^2 - 3x_1x_2 + x_2^2$
and $\mathbf{R} = \langle u_1 + 2u_2, u_1 + u_2 \rangle$

b) $f(x_1, x_2, x_3) = x_1^2 + 2x_2^2 + 3x_3^2 - 2x_1x_2 + 4x_2x_3$
and $\mathbf{R} = \langle u_1 + u_2 - u_3, -u_1 + 2u_2 + u_3, 2u_1 + 2u_2 - u_3 \rangle$

c) $f(x_1, x_2) = x_1^2 + 5x_1x_2 + 3x_2^2$
and $\mathbf{R} = \langle u_1 - 2u_2 + u_3, -u_1 + u_2 - 3u_3, 2u_1 - u_2 + u_3 \rangle$.

4. If $f = f(u)$ and $u = u(x_1, \ldots, x_n)$ then $g = g(x_1, \ldots, x_n) = f(u(x_1, \ldots, x_n))$. Verify the chain rule

$$\left\langle \frac{\partial g}{\partial x_1}, \ldots, \frac{\partial g}{\partial x_n} \right\rangle = f'(u) \left\langle \frac{\partial u}{\partial x_1}, \ldots, \frac{\partial u}{\partial x_n} \right\rangle$$

if

a) $f(u) = e^u$ and $u = x^3 y$

b) $f(u) = \ln u$ and $u = x^3 + y + z^2$

c) $f(u) = u^{-1}$ and $u = \sqrt{x^2 + y^2 + z^2}$

d) $f(u) = \cos u$ and $u = 2x + y^2 + z^3$.

5. If $f = f(x, y)$ and $\mathbf{R} = \langle x, y \rangle = \mathbf{R}(t)$ then $g = g(t) = f(x(t), y(t))$. Verify the chain rule

$$\frac{dg}{dt} = \begin{pmatrix} \frac{\partial f}{\partial x} & \frac{\partial f}{\partial y} \end{pmatrix} \begin{pmatrix} \frac{dx}{dt} \\ \frac{dy}{dt} \end{pmatrix}$$

if

a) $f(x, y) = x^2 - 3xy + 2y^2$ and $\mathbf{R} = \langle 3t, t \rangle$

b) $f(x, y) = 2x^3 - 5xy$ and $\mathbf{R} = \langle t, t^2 \rangle$

c) $f(x, y) = 3x^5 - y^3$ and $\mathbf{R} = \langle t^3, t^5 \rangle$

d) $f(x, y) = x^2 + y^2$ and $\mathbf{R} = \langle t^2 \cos t, t^2 \sin t \rangle$.

6. If $f = f(x, y)$ and if a mapping from $\mathfrak{R}^2$ to $\mathfrak{R}^2$ is given by $\mathbf{R} = \langle x, y \rangle = \mathbf{R}(u, v)$ then $g = g(u, v) = f(x(u, v), y(u, v))$. Verify the chain rule

$$\begin{pmatrix} \frac{\partial g}{\partial u} & \frac{\partial g}{\partial v} \end{pmatrix} = \begin{pmatrix} \frac{\partial f}{\partial x} & \frac{\partial f}{\partial y} \end{pmatrix} \begin{pmatrix} \frac{\partial x}{\partial u} & \frac{\partial x}{\partial v} \\ \frac{\partial y}{\partial u} & \frac{\partial y}{\partial v} \end{pmatrix}$$

if

a) $f(x, y) = x^2 + y^2$ and $\mathbf{R} = \langle u \cos v, u \sin v \rangle$

b) $f(x, y) = x^2 - y^2$ and $\mathbf{R} = \langle u \tan v, u \sec v \rangle$

c) $f(x, y) = 3x + 2y$ and $\mathbf{R} = \langle u + 5v, u + 2v \rangle$

d) $f(x, y) = xy$ and $\mathbf{R} = \langle u + 2v, u - 2v \rangle$.

7. If $f = f(x, y, z)$ and if $\mathbf{R} = \langle x, y, z \rangle = \mathbf{R}(t)$ then $g = g(t) = f(x(t), y(t), z(t))$. Verify the chain rule

$$\frac{dy}{dt} = \begin{pmatrix} \frac{\partial f}{\partial x} & \frac{\partial f}{\partial y} & \frac{\partial f}{\partial z} \end{pmatrix} \begin{pmatrix} \frac{dx}{dt} \\ \frac{dy}{dt} \\ \frac{dz}{dt} \end{pmatrix}$$

if

a) $f(x,y,z) = 2xy - z$ and $\mathbf{R} = \langle t, t^2, t^3 \rangle$

b) $f(x,y,z) = 2x^2y - z$ and $\mathbf{R} = \langle t, t^2, t^4 \rangle$

c) $f(x,y,z) = x^2 + y^2 - 2z$ and $\mathbf{R} = \langle t\cos t, t\sin t, t^2 \rangle$

d) $f(x,y,z) = x^2 - y^2 + z$ and $\mathbf{R} = \langle t\tan t, t\sec t, t^2 \rangle$.

8. If $f(x,y,z)$ and if $\mathbf{R} = \langle x,y,z \rangle = \mathbf{R}(u,v)$ then

$$g = g(u,v) = f(x(u,v), y(u,v), z(u,v))) \quad .$$

Verify the chain rule

$$\begin{pmatrix} \frac{\partial g}{\partial u} & \frac{\partial g}{\partial v} \end{pmatrix} = \begin{pmatrix} \frac{\partial f}{\partial x} & \frac{\partial f}{\partial y} & \frac{\partial f}{\partial z} \end{pmatrix} \begin{pmatrix} \frac{\partial x}{\partial u} & \frac{\partial x}{\partial v} \\ \frac{\partial y}{\partial u} & \frac{\partial y}{\partial v} \\ \frac{\partial z}{\partial u} & \frac{\partial z}{\partial v} \end{pmatrix}$$

if

a) $f(x,y,z) = xy - z^2$ and $\mathbf{R} = \langle (u+v)^2, \tan^2 v, (u-v)\sec v \rangle$

b) $f(x,y,z) = 2xy - z^2$ and $\mathbf{R} = \langle u\cos v, u\sin v, u \rangle$

c) $f(x,y,z) = x^2 - y^2 - 2z^2$ and $\mathbf{R} = \langle u\cosh v, u\sinh v, u \rangle$.

9. If $f = f(x,y,z)$ and if a mapping from $\mathcal{R}^3$ to $\mathcal{R}^3$ is given $\mathbf{R} = \langle x,y,z \rangle = \mathbf{R}(u,v,w)$ then $g = g(u,v,w) = f(x(u,v,w), y(u,v,w), z(u,v,w))$. Verify the chain rule

$$\begin{pmatrix} \frac{\partial g}{\partial u} & \frac{\partial g}{\partial v} & \frac{\partial g}{\partial w} \end{pmatrix}$$

$$= \begin{pmatrix} \frac{\partial f}{\partial x} & \frac{\partial f}{\partial y} & \frac{\partial f}{\partial z} \end{pmatrix} \begin{pmatrix} \frac{\partial x}{\partial u} & \frac{\partial x}{\partial v} & \frac{\partial x}{\partial w} \\ \frac{\partial y}{\partial u} & \frac{\partial y}{\partial v} & \frac{\partial y}{\partial w} \\ \frac{\partial z}{\partial u} & \frac{\partial z}{\partial v} & \frac{\partial z}{\partial w} \end{pmatrix}$$

if

a) $f(x,y,z) = 2x + 3y - 5z$ and $\mathbf{R} = \langle u + v + 2w, -u + v + w, 2u + v - w\rangle$

b) $f(x,y,z) = x^2 + y^2 - z^2$ and $\mathbf{R} = \langle u\cos v, u\sin v, w\rangle$

c) $f(x,y,z) = x^2 - y^2 - z^2$ and $\mathbf{R} = \langle -u\cosh v, u\sinh v, w\rangle$

d) $f(x,y,z) = x^2 + y^2 + z^2$ and $\mathbf{R} = \langle u\cos v, u\sin v, w\rangle$.

10. If $f = f(x,y)$ and if $\mathbf{R} = \langle x, y\rangle = \mathbf{R}(u,v,w)$ then

$$g = g(u,v,w) = f(x(u,v,w), y(u,v,w)) \quad .$$

Verify the chain rule

$$\begin{pmatrix} \frac{\partial g}{\partial u} & \frac{\partial g}{\partial v} & \frac{\partial g}{\partial w} \end{pmatrix} = \begin{pmatrix} \frac{\partial f}{\partial x} & \frac{\partial f}{\partial y} \end{pmatrix} \begin{pmatrix} \frac{\partial x}{\partial u} & \frac{\partial x}{\partial v} & \frac{\partial x}{\partial w} \\ \frac{\partial y}{\partial u} & \frac{\partial y}{\partial v} & \frac{\partial y}{\partial w} \end{pmatrix}$$

if

a) $f(x,y) = 2x - y$ and $\mathbf{R} = \langle u + 2v - w, -u + v + 3w\rangle$

b) $f(x,y) = x^2 + 3xy - y^2$ and $\mathbf{R} = \langle u - v + 2w, 2u + 3v - w\rangle$

c) $f(x,y) = -x^2 + xy + 2y^2$ and $\mathbf{R} = \langle u + v + 2w, u - v - w\rangle$.

11 Linear and Quadratic Approximations

Outline of Key Ideas

Linear and quadratic approximations for $y = \phi(x)$.

Linear and quadratic approximations for $f = f(x_1, \ldots, x_n)$:

$$\mathbf{L}f(\mathbf{x}) = f(\mathbf{a}) + \nabla f(\mathbf{a})(\mathbf{x} - \mathbf{a})$$

$$\mathbf{Q}f(\mathbf{x}) = \mathbf{L}f(\mathbf{x}) + \tfrac{1}{2}(\mathbf{x} - \mathbf{a})^T \mathbf{H}f(\mathbf{a})(\mathbf{x} - \mathbf{a})$$

$$\mathbf{H}f(\mathbf{a}) = \left[\frac{\partial^2 f(\mathbf{a})}{\partial x_i \partial x_j}\right]_{n \times n} = \text{Hessian matrix of } f \text{ at } \mathbf{a}.$$

Linear and quadratic approximations for $f(x_1, x_2)$ and $f(x_1, x_2, x_3)$.

Linear and Quadratic Approximations

11.1 Linear and quadratic approximations for $y = \phi(x)$

If $\phi = \phi(x)$ is a real valued function and if $\phi''(t)$ exists for t between a and x then by Taylor's Formula with Remainder we have

$$\phi(x) = \phi(a) + \frac{\phi'(a)}{1!}(x-a) + \frac{\phi''(t_0)}{2!}(x-a)^2$$

for some t_0 between a and x. Then

$$\mathbf{L}\phi(x) = \phi(a) + \phi'(a)(x-a) \tag{11.1}$$

is the ***linear approximation of*** $\phi(x)$ ***about*** a. The expression $y = \mathbf{L}\phi(x)$ is just the equation of the tangent line at a to the graph of $y = \phi(x)$. If $\phi'''(t)$ exists for t between a and x then once again by Taylor's Formula with Remainder we have

$$\phi(x) = \phi(a) + \frac{\phi'(a)}{1!}(x-a) + \frac{\phi''(a)}{2!}(x-a)^2 + \frac{\phi'''(t_0)}{3!}(x-a)^3$$

for some t_0 between a and x. Then

$$\mathbf{Q}\phi(x) = \phi(a) + \phi'(a)(x-a) + \frac{1}{2}\phi''(a)(x-a)^2 \tag{11.2}$$

is the ***quadratic approximation of*** $\phi(x)$ ***about*** a. From the linear approximation we can obtain some information about $\phi(x)$ near a. For instance if $\phi'(a)$ were positive then near a we would have $\phi(x) < \phi(a)$ for $x < a$ and $\phi(a) < \phi(x)$ for $a < x$. Thus $\phi(x)$ would be increasing at a. From the quadratic approximation we can obtain more information about $\phi(x)$ near a. For instance if $\phi''(a)$ were negative, then near a $\phi(x) - \mathbf{L}\phi(x)$ would be negative or we would have $\phi(x) \leq \mathbf{L}\phi(x)$, i.e. near a the graph of $y = \phi(x)$ would lie under the tangent line to the graph at a.

Example 11.1

If $\phi(x) = \arctan x$ then $\phi'(x) = \dfrac{1}{1+x^2}$ and $\phi''(x) = \dfrac{-2x}{(1+x^2)^2}$. Thus, $\phi(1) = \dfrac{\pi}{4}$, $\phi'(1) = \dfrac{1}{2}$, and $\phi''(1) = \dfrac{-1}{2}$. So the linear approximation $\mathbf{L}\phi(x)$ is given by $\mathbf{L}\phi(x) = \phi(1) + \phi'(1)(x-1) =$

$$\mathbf{L}\phi(x) = \frac{\pi}{4} + \frac{1}{2}(x-1)$$

and the quadratic approximation is given by $\mathbf{Q}\phi(x) = \phi(1) + \phi'(1)(x-1) + \frac{1}{2}\phi''(1)(x-1)^2 =$

$$\mathbf{Q}\phi(x) = \frac{\pi}{4} + \frac{1}{2}(x-1) - \frac{1}{4}(x-1)^2 \quad .$$

11.2 Linear and Quadratic Approximations for $f = f(x_1, \ldots, x_n)$

If $f : \mathfrak{R}^n \to \mathfrak{R}$ is a real valued function of n variables we can use matrices to represent its linear and quadratic approximations near a fixed point. The problem of defining and finding the linear and quadratic approximation is reduced to the one variable case by only considering the points on a straight line segment which depend on a single parameter t with $0 \le t \le 1$.

Let each point in $\mathfrak{R}^n$ be represented by its ***position vector*** $\mathbf{x} = \langle x_1, \ldots, x_n \rangle$ then $f(x_1, \ldots, x_n) = f(\mathbf{x})$. In order to find linear and quadratic approximations to $f(\mathbf{x})$ near a point $\mathbf{a} = \langle a_1, \ldots, a_n \rangle$ we will restrict $f(\mathbf{x})$ to the points $\mathbf{y} = t(\mathbf{x} - \mathbf{a}) + \mathbf{a}$ on the line segment connecting $\mathbf{a}$ and $\mathbf{x}$. Then $f(\mathbf{y}) = f(t(\mathbf{x} - \mathbf{a}) + \mathbf{a}) = \phi(t)$ is a function of t with $0 \le t \le 1$. The linear approximation to $\phi(t)$ near 0 is given by

$$\phi(0) + \phi'(0)t$$

and the quadratic approximation to $\phi(t)$ near 0 is given by

$$\phi(0) + \phi'(0)t + \tfrac{1}{2}\phi''(0)t \quad .$$

For $t = 0$ $\mathbf{y}(0) = \mathbf{a}$ and for $t = 1$ $\mathbf{y}(1) = \mathbf{x}$. Thus $\mathbf{L}\phi(1)$ is given by $\phi(0) + \phi'(0)$ and $\mathbf{Q}\phi(1)$ is given by $\phi(0) + \phi'(0) + \frac{1}{2}\phi''(0)$ after we set $t = 1$. It remains to find $\phi(0)$, $\phi'(0)$, and $\phi''(0)$. The chain rule is needed on $\phi(t) = f(\mathbf{y}(t))$.

Let $f_k(\mathbf{a})$ denote the partial derivative of f and $\mathbf{a}$ with respect to the kth variable, i.e. $f_k(\mathbf{a}) = \dfrac{\partial f}{\partial x_k}$ at $\mathbf{a}$. Now by the chain rule

$$\phi'(t) = \sum_{i=1}^{n} f_i(\mathbf{y})\frac{dy_i}{dt} = \sum_{i=1}^{n} f_i(t(\mathbf{x} - \mathbf{a}) + \mathbf{a})(x_i - a_i)$$

where $y_i = (x_i - a_i)t + a_i$ and $\dfrac{dy_i}{dt} = x_i - a_i$ for $i = 1, 2 \ldots, n$. But then

$$(11.3) \qquad \phi'(0) = \sum_{i=1}^{n} f_i(\mathbf{a})(x_i - a_i) = \nabla f\Big|_{\mathbf{a}} \bullet (\mathbf{x} - \mathbf{a})$$

$$= \begin{pmatrix} f_1(\mathbf{a}) & \cdots & f_n(\mathbf{a}) \end{pmatrix} \begin{pmatrix} x_1 - a_1 \\ \vdots \\ x_n - a_n \end{pmatrix} .$$

Repeat the process with $\phi'(t)$ in place of $\phi(t)$ to obtain

$$\phi''(t) = \sum_{i=1}^{n} \sum_{j=1}^{n} f_{ij}(t(\mathbf{x} - \mathbf{a}) + \mathbf{a})(x_i - a_i)(x_j - a_j) \quad .$$

Then

$$\phi''(0) = \sum_{i=1}^{n} \sum_{j=1}^{n} (x_i - a_i) f_{ij}(\mathbf{a})(x_j - a_j)$$

$$(11.4) \qquad = \begin{pmatrix} x_1 - a_1, & \ldots, & x_n - a_n \end{pmatrix} \begin{pmatrix} f_{11}(\mathbf{a}) & \cdots & f_{1n}(\mathbf{a}) \\ \vdots & & \vdots \\ f_{n1}(\mathbf{a}) & \ldots & f_{nn}(\mathbf{a}) \end{pmatrix} \begin{pmatrix} x_1 - a_1 \\ \vdots \\ x_n - a_n \end{pmatrix}$$

$$= (\mathbf{x} - \mathbf{a})^T \mathbf{H} f(\mathbf{a})(\mathbf{x} - \mathbf{a})$$

where $\mathbf{H}f(\mathbf{a})$ is the $n \times n$ ***Hessian matrix*** of f at $\mathbf{a}$, i.e. $[f_{ij}(\mathbf{a})]_{n \times n}$ the matrix of second partial derivatives of f at $\mathbf{a}$.

Thus,

$$(11.5) \qquad \mathbf{L}f(\mathbf{x}) = f(\mathbf{a}) + \nabla f\Big|_{\mathbf{a}} \bullet (\mathbf{x} - \mathbf{a})$$

$$= f(\mathbf{a}) + \begin{pmatrix} f_1(\mathbf{a}), & \ldots, & f_n(\mathbf{a}) \end{pmatrix} \begin{pmatrix} x_1 - a_1 \\ \vdots \\ x_n - a_n \end{pmatrix}$$

is the ***linear approximation*** of $f(\mathbf{x})$ about $\mathbf{a}$. For $n = 2$ the expression $z = \mathbf{L}f(x, y)$ is just the equation of the tangent plane to the graph of $z = f(x, y)$. Likewise,

$$\mathbf{Q}f(\mathbf{x}) = f(\mathbf{a}) + \nabla f\Big|_{\mathbf{a}} \bullet (\mathbf{x} - \mathbf{a}) + \frac{1}{2}(\mathbf{x} - \mathbf{a})^T \mathbf{H} f(\mathbf{a})(\mathbf{x} - \mathbf{a})$$

$$(11.6) \qquad = f(\mathbf{a}) + \begin{pmatrix} f_1(\mathbf{a}), & \ldots, & f_n(\mathbf{a}) \end{pmatrix} \begin{pmatrix} x_1 - a_1 \\ \vdots \\ x_n - a_n \end{pmatrix}$$

$$+\frac{1}{2}\begin{pmatrix} x_1-a_1, & \ldots, & x_n-a_n \end{pmatrix}\begin{pmatrix} f_{11}(\mathbf{a}) & \cdots & f_{1n}(\mathbf{a}) \\ \vdots & & \vdots \\ f_{n1}(\mathbf{a}) & \cdots & f_{nn}(\mathbf{a}) \end{pmatrix}\begin{pmatrix} x_1-a_1 \\ \vdots \\ x_n-a_n \end{pmatrix}$$

is the ***quadratic approximation*** of $f(\mathbf{x})$ about $\mathbf{a}$. When all of the second partial derivatives f_{ij} are continuous for $i \neq j$ then $\mathbf{H}f(\mathbf{a})$ is a symmetric matrix and the last expression in $\mathbf{Q}f(\mathbf{x})$ is a quadratic form $\frac{1}{2}(\mathbf{x}-\mathbf{a})^T\mathbf{H}f(\mathbf{a})(\mathbf{x}-\mathbf{a})$. We studied such quadratic forms in Chapter 8. In the next chapter we will see that the eigenvalues of $\mathbf{H}f(\mathbf{a})$ at a critical point of f will be very useful in determining the nature of the critical point.

11.3 Linear and Quadratic Approximations for $f = f(x_1, x_2)$ and $f = f(x_1, x_2, x_3)$

For a function of two variables $f = f(x_1, x_2)$ its ***linear approximation*** $\mathbf{L}f(x_1, x_2)$ about $\mathbf{a} = (a_1, a_2)$ is given by

$$\mathbf{L}f(x_1,x_2) = f(a_1,a_2) + f_1(a_1,a_2)(x_1-a_1) + f_2(a_1,a_2)(x_2-a_2) \tag{11.7}$$

where $f_1(a_1,a_2) = \left.\dfrac{\partial f}{\partial x_1}\right|_{(a_1,a_2)}$ and $f_2(a_1,a_2) = \left.\dfrac{\partial f}{\partial x_2}\right|_{(a_1,a_2)}$.

Its ***quadratic approximation*** $\mathbf{Q}f(x_1, x_2)$ about $\mathbf{a} = (a_1, a_2)$ is given by

$$\mathbf{Q}f(x_1,x_2) = f(a_1,a_2) + f_1(a_1,a_2)(x_1-a_1) + f_2(a_1,a_2)(x_2-a_2) \tag{11.8}$$

$$+\frac{1}{2}\begin{pmatrix} x_1-a_1, & x_2-a_2 \end{pmatrix}\begin{pmatrix} f_{11}(a_1,a_2) & f_{12}(a_1,a_2) \\ f_{12}(a_1,a_2) & f_{22}(a_1,a_2) \end{pmatrix}\begin{pmatrix} x_1-a_1 \\ x_2-a_2 \end{pmatrix}$$

$$= f(a_1,a_2) + f_1(a_1,a_2)(x_1-a_1) + f_2(a_1,a_2)(x_2-a_2)$$

$$+\frac{1}{2}\left[f_{11}(a_1,a_2)(x_1-a_1)^2 + 2f_{12}(a_1,a_2)(x_1-a_1)(x_2-a_2) + f_{22}(x_2-a_2)^2\right]$$

where

$$f_{ij}(a_1,a_2) = \left.\frac{\partial^2 f}{\partial x_i \partial x_j}\right|_{(a_1,a_2)} \qquad i,j = 1,2$$

are the second partial derivatives of f at $\mathbf{a} = (a_1, a_2)$. The ***Hessian matrix*** $\mathbf{H}f(\mathbf{a})$ of second partial derivatives is symmetric when the $f_{12}(a_1, a_2)$ and $f_{21}(a_1, a_2)$ are continuous which we will assume. Observe that

$$\mathbf{H}f(a_1, a_2) = \left(\begin{array}{cc} \dfrac{\partial}{\partial x_1}\left(\dfrac{\partial f}{\partial x_1}\right) & \dfrac{\partial}{\partial x_2}\left(\dfrac{\partial f}{\partial x_1}\right) \\ \dfrac{\partial}{\partial x_1}\left(\dfrac{\partial f}{\partial x_2}\right) & \dfrac{\partial}{\partial x_2}\left(\dfrac{\partial f}{\partial x_2}\right) \end{array} \right)_{(a_1, a_2)} \tag{11.9}$$

$$= \left(\begin{array}{c} \nabla f_1 \\ \nabla f_2 \end{array} \right)_{(a_1, a_2)} , \tag{11.10}$$

i.e. the first row of $\mathbf{H}f$ is the gradient of f_1 and the second row of $\mathbf{H}f$ is the gradient of f_2.

Example 11.2

Find the quadratic approximation to $f(x, y) = \ln(1 + 2x + 3y)$ at
a) $(a_1, a_2) = (0, 0)$ and **b)** $(a_1, a_2) = (-1, 1)$.

Solution.

In either case we need f_1, f_2, f_{11}, f_{12}, and f_{22} where $x = x_1$ is the first variable and $y = x_2$ is the second variable. Thus $f_1 = f_x = \dfrac{2}{1 + 2x + 3y}$ and $f_2 = f_y = \dfrac{3}{1 + 2x + 3y}$. Now, $\langle f_{11}, f_{12} \rangle = \nabla f_x = \langle f_{xx}, f_{xy} \rangle = \left(\dfrac{1}{(1 + 2x + 3y)^2} \right) \langle -4, -6 \rangle$ and $\langle f_{21}, f_{22} \rangle = \nabla f_y = \langle f_{yx}, f_{yy} \rangle = \left(\dfrac{1}{(1 + 2x + 3y)^2} \right) \langle -6, -9 \rangle$, so

$$\nabla f = \frac{1}{1 + 2x + 3y} \langle 2, 3 \rangle$$

and

$$\mathbf{H}f = \frac{1}{(1 + 2x + 3y)^2} \left(\begin{array}{cc} -4 & -6 \\ -6 & -9 \end{array} \right) .$$

Then for Part (a) we have the following:

$$f(0, 0) = \ln 1 = 0$$

$$\left.\nabla f\right|_{(0,0)} = \langle 2,3\rangle$$
$$\mathbf{H}f(0,0) = \begin{pmatrix} -4 & -6 \\ -6 & -9 \end{pmatrix} .$$

Thus,

$$\begin{aligned}\mathbf{Q}f(x,y) &= 0+2(x-0)+3(y-0) \\ &\quad +\frac{1}{2}\begin{pmatrix} x-0, & y-0 \end{pmatrix}\begin{pmatrix} -4 & -6 \\ -6 & -9 \end{pmatrix}\begin{pmatrix} x-0 \\ y-0 \end{pmatrix} \\ &= 2x+3y-\frac{1}{2}(4x^2+12xy+9y^2)\end{aligned}$$

is the quadratic approximation to f about $(0,0)$. For Part (b) we have the following:

$$f(-1,1) = \ln 2$$
$$\left.\nabla f\right|_{(-1,1)} = \frac{1}{2}\langle 2,3\rangle$$
$$\mathbf{H}f(-1,1) = \frac{1}{4}\begin{pmatrix} -4 & -6 \\ -6 & -9 \end{pmatrix} .$$

Thus,

$$\mathbf{Q}f(x,y) = \ln 2+\frac{1}{2}(2(x+1)+3(y-1))$$

$$+\frac{1}{2}\begin{pmatrix} x+1, & y-1 \end{pmatrix}\frac{1}{4}\begin{pmatrix} -4 & -6 \\ -6 & -9 \end{pmatrix}\begin{pmatrix} x+1 \\ y-1 \end{pmatrix}$$

$$= \ln 2+\frac{1}{2}(2(x+1)+3(y-1))-\frac{1}{8}(4(x+1)^2+12(x+1)(y-1)+9(y-1)^2)$$

is the quadratic approximation to f about $(-1,1)$. It can be simplified to

$$\mathbf{Q}f(x,y) = \ln 2+(x+1)+\frac{3}{2}(y-1)-\frac{1}{2}(x+1)^2-\frac{3}{2}(x+1)(y-1)-\frac{9}{8}(y-1)^2 .$$

In this form you can identify it as a quadratic approximation to f about $(-1,1)$.

Warning: Do not attempt any further "simplification" or else you will lose sight of the fact that this is an approximation about $(-1,1)$. For instance if all of the above terms were multiplied out we would obtain the following misleading expression

$$\mathbf{Q}f(x,y) = (\ln 2 - \tfrac{5}{8}) + \tfrac{3}{2}x + \tfrac{9}{4}y - \tfrac{1}{2}x^2 - \tfrac{3}{2}xy - \tfrac{9}{8}y^2$$

which appears to be the quadratic approximation to f about $(0,0)$, but it is not! In this form there is no evidence that the point $(-1,1)$ is involved. It is essential to leave linear and quadratic approximations in terms of the differences $x_i - a_i$, $i = 1,2$.

For a function of three variables its ***linear approximation*** $\mathbf{L}f(x_1,x_2,x_3)$ about $\mathbf{a} = (a_1,a_2,a_3)$ is given by

$$(11.11)\quad \begin{aligned}\mathbf{L}f(x_1,x_2,x_3) &= f(\mathbf{a}) + \begin{pmatrix} f_1(\mathbf{a}) & f_2(\mathbf{a}) & f_3(\mathbf{a}) \end{pmatrix} \begin{pmatrix} x_1 - a_1 \\ x_2 - a_2 \\ x_3 - a_3 \end{pmatrix} \\ &= f(a_1,a_2,a_3) + f_1(a_1,a_2,a_3)(x_1 - a_1) \\ &\quad + f_2(a_1,a_2,a_3)(x_2 - a_2) + f_3(a_1,a_2,a_3)(x_3 - a_3) \quad .\end{aligned}$$

Its ***quadratic approximation*** $\mathbf{Q}f(x_1,x_2,x_3)$ about $\mathbf{a} = (a_1,a_2,a_3)$ is given by

$$(11.12)\qquad \mathbf{Q}f(x_1,x_2,x_3) = f(\mathbf{a}) + \begin{pmatrix} f_1(\mathbf{a}) & f_2(\mathbf{a}) & f_3(\mathbf{a}) \end{pmatrix} \begin{pmatrix} x_1 - a_1 \\ x_2 - a_2 \\ x_3 - a_3 \end{pmatrix}$$

$$+\frac{1}{2} \begin{pmatrix} x_1 - a_1, & x_2 - a_2, & x_3 - a_3 \end{pmatrix} \begin{pmatrix} f_{11}(\mathbf{a}) & f_{12}(\mathbf{a}) & f_{13}(\mathbf{a}) \\ f_{12}(\mathbf{a}) & f_{22}(\mathbf{a}) & f_{23}(\mathbf{a}) \\ f_{13}(\mathbf{a}) & f_{23}(\mathbf{a}) & f_{33}(\mathbf{a}) \end{pmatrix} \begin{pmatrix} x_1 - a_1 \\ x_2 - a_2 \\ x_3 - a_3 \end{pmatrix}$$

$$= f(\mathbf{a}) + f_1(\mathbf{a})(x_1 - a_1) + f_2(\mathbf{a})(x_2 - a_2) + f_3(\mathbf{a})(x_3 - a_3)$$

$$+\frac{1}{2}\Big[f_{11}(\mathbf{a})(x_1 - a_1)^2 + f_{22}(\mathbf{a})(x_2 - a_2)^2 + f_{33}(\mathbf{a})(x_3 - a_3)^2$$

$$+2f_{12}(\mathbf{a})(x_1 - a_1)(x_2 - a_2) + 2f_{13}(\mathbf{a})(x_1 - a_1)(x_3 - a_3) + 2f_{23}(\mathbf{a})(x_2 - a_2)(x_3 - a_3)\Big] \quad .$$

Again it is essential to leave $\mathbf{L}f$ and $\mathbf{Q}f$ in terms of the differences $x_i - a_i$, $i = 1,2,3$.

Example 11.3 If $f(x,y) = 35 + 2x - 3y + 5x^2 + 4xy + 3y^2$ find $f(0,0)$, $\nabla f(0,0)$, and $\mathbf{H}f(0,0)$.

Solution.

Write $f(x,y)$ as

$$f(x,y) = 35 + \begin{pmatrix} 2 & -3 \end{pmatrix} \begin{pmatrix} x \\ y \end{pmatrix} + \begin{pmatrix} x & y \end{pmatrix} \begin{pmatrix} 5 & 2 \\ 2 & 3 \end{pmatrix} \begin{pmatrix} x \\ y \end{pmatrix}$$

and observe that

$$f(x,y) = f(0,0) + \nabla f(0,0) \begin{pmatrix} x \\ y \end{pmatrix} + \frac{1}{2} \begin{pmatrix} x & y \end{pmatrix} \mathbf{H}f(0,0) \begin{pmatrix} x \\ y \end{pmatrix} .$$

It follows that $f(0,0) = 35$, $\nabla f(0,0) = \langle 2, -3 \rangle$, and

$$\mathbf{H}f(0,0) = \begin{pmatrix} 10 & 4 \\ 4 & 6 \end{pmatrix} .$$

Example 11.4 If $\mathbf{a} = (1,2,3)$, $f(\mathbf{a}) = 4$, $\nabla f(\mathbf{a}) = \langle 5,6,7 \rangle$, and

$$\mathbf{H}f(\mathbf{a}) = \begin{pmatrix} 9 & 8 & 6 \\ 8 & 7 & 10 \\ 6 & 10 & 5 \end{pmatrix}$$

find $\mathbf{Q}f$ the quadratic approximation to $f(x_1, x_2, x_3)$ about $\mathbf{a}$.

Solution.

$$\mathbf{Q}f(x_1,x_2,x_3) = 4 + \begin{pmatrix} 5 & 6 & 7 \end{pmatrix} \begin{pmatrix} x-1 \\ y-2 \\ z-3 \end{pmatrix}$$

$$+\frac{1}{2} \begin{pmatrix} x-1, & y-2, & z-3 \end{pmatrix} \begin{pmatrix} 9 & 8 & 6 \\ 8 & 7 & 10 \\ 6 & 10 & 5 \end{pmatrix} \begin{pmatrix} x-1 \\ y-2 \\ z-3 \end{pmatrix}$$

$$= 4 + 5(x-1) + 6(y-2) + 7(z-3) + \frac{1}{2}\Big[9(x-1)^2 + 7(y-2)^2 + 5(z-3)^2$$

$$+16(x-1)(y-2)+12(x-1)(z-3)+20(y-2)(z-3)\Big]$$

$$=4+5(x-1)+6(y-2)+7(z-3)+\frac{9}{2}(x-1)^2+\frac{7}{2}(y-2)^2+\frac{5}{2}(z-3)^2$$

$$+8(x-1)(y-2)+6(x-1)(z-3)+10(y-2)(z-3) \quad .$$

Observe that

$$\mathbf{Q}f(\mathbf{x})=\mathbf{L}f(\mathbf{x})+\frac{1}{2}(\mathbf{x}-\mathbf{a})^T\mathbf{H}f(\mathbf{a})(\mathbf{x}-\mathbf{a}) \quad ,$$

i.e. the quadratic approximation to f about $\mathbf{a}$ equals the linear approximation to f about $\mathbf{a}$ plus the additional quadratic form $\frac{1}{2}(\mathbf{x}-\mathbf{a})^T\mathbf{H}f(\mathbf{a})(\mathbf{x}-\mathbf{a})$ centered at $\mathbf{a}$. Thus, $\mathbf{L}f$ is always part of $\mathbf{Q}f$, so the homework exercises will only involve $\mathbf{Q}f$ explicitly.

11.4 Exercises

1. Find the quadratic approximation to $f(x,y)=\dfrac{1+y}{x}$ about the point

a) $(1,2)$ **b)** $(1,-1)$ **c)** $(3,2)$ **d)** $(3,-1)$.

2. If $f(x,y)=361-19x+18y+13x^2+3xy+7y^2+x^3+x^2y-3xy^2-y^3$ find

a) $f(0,0)$, $\nabla f(0,0)$, $\mathbf{H}f(0,0)$
b) $f(1,1)$, $\nabla f(1,1)$, $\mathbf{H}f(1,1)$
c) $f(1,-1)$, $\nabla f(1,-1)$, $\mathbf{H}f(1,-1)$
d) $f(-1,1)$, $\nabla f(-1,1)$, $\mathbf{H}f(-1,1)$.

3. If

$$\begin{aligned} f(x,y) &= 121+9(x+2)+7(y-3)+5(x+2)^2+4(x+2)(y-3)-3(y-3)^2 \\ &\quad +(x+2)^3-6(x+2)^2(y-3)+15(x+2)(y-3)^2+3(y-3)^3 \end{aligned}$$

find

a) $f(-2,3)$, $\nabla f(-2,3)$, and $\mathbf{H}f(-2,3)$

b) $f(0,0)$, $\nabla f(0,0)$, and $\mathbf{H}f(0,0)$.

4. Find the quadratic approximation $\mathbf{Q}f(x,y,z)$ to $f(x,y,z) = \dfrac{x-y}{z}$ about

a) $\mathbf{a} = (3,2,1)$ **b)** $\mathbf{a} = (2,3,1)$ **c)** $\mathbf{a} = (1,2,-1)$ **d)** $\mathbf{a} = (1,-1,1)$.

5. If $f(x,y,z) = xy^2+x^2z+y^4z^3$ find $\nabla f(\mathbf{a})$, $\mathbf{H}f(\mathbf{a})$, and the quadratic approximation $\mathbf{Q}f(x,y,z)$ to $f(x,y,z)$ about

a) $\mathbf{a} = (-1,1,-1)$ **b)** $\mathbf{a} = (-1,1,1)$ **c)** $\mathbf{a} = (1,-1,-1)$ **d)** $\mathbf{a} = (1,1,-1)$.

6. Find the quadratic approximation $\mathbf{Q}f(x,y,z)$ to $f(x,y,z)$ at $\mathbf{a}$ if

a) $f(x,y,z) = \ln(x^2+2y+2z)$ and $\mathbf{a} = (1,1,-1)$

b) $f(x,y,z) = 2z\arctan\left(\dfrac{x}{y}\right)$ and $\mathbf{a} = (1,-1,1)$

c) $f(x,y,z) = x^5+xy^3-3z^5$ and $\mathbf{a} = (1,-1,1)$

d) $f(x,y,z) = 2x+3y-6z+2yz-5xyz$ and $\mathbf{a} = (1,-1,-1)$.

7. If $f(x,y,z) = 3+[1,-1,1]A\mathbf{r}+\mathbf{r}^T B\mathbf{r}$ where $\mathbf{r} = [x,y,z]^T$,

$$A = \begin{pmatrix} 1 & 1 & 1 \\ 2 & 1 & 2 \\ -2 & 2 & 1 \end{pmatrix} \quad \text{and} \quad B = \begin{pmatrix} 2 & 8 & 7 \\ 0 & 3 & 1 \\ 3 & 5 & -5 \end{pmatrix}$$

find

a) $\nabla f(0,0,0)$ and $\mathbf{H}f(0,0,0)$

b) $\mathbf{Q}f(x,y,z)$ the quadratic approximation to $f(x,y,z)$ about $(0,0,0)$.

8. If $f(x,y,z) = 5+[-1,1,1]A\mathbf{r}+\mathbf{r}^T B\mathbf{r}$ where $\mathbf{r} = [x,y,z]^T$

$$A = \begin{pmatrix} 1 & 2 & 1 \\ 2 & 1 & 3 \\ 3 & 1 & -3 \end{pmatrix} \quad \text{and} \quad B = \begin{pmatrix} 3 & 0 & 13 \\ 16 & 1 & 5 \\ 1 & 3 & 7 \end{pmatrix}$$

find

a) $\nabla f(0,0,0)$ and $\mathbf{H}f(0,0,0)$

b) $\mathbf{Q}f(x,y,z)$ the quadratic approximation to $f(x,y,z)$ about $(0,0,0)$.

12 Maxima, Minima, and Saddle Points for Functions of Several Variables

Outline of Key Ideas

Basic concepts.

Critical points and quadratic approximations at critical points.

Analysis of the behavior of a function at its critical points.

Maxima, Minima, and Saddle Points for Functions of Several Variables

12.1 Basic Concepts

Derivatives are used for functions of one variable to determine their critical points and to determine where they have local maxima, local minima, and points of inflection.

Similarly, for functions of several variables partial derivatives are used to determine their critical points and to determine where they have local maxima, local minima, and saddle points. Quadratic approximations and the eigenvalues of the Hessian matrix are used to analyze these critical points.

Before we start some vocabulary needs to be introduced.

In $\mathcal{R}^2$ an open disk with center at $\mathbf{a} = (a_1, a_2)$ and positive radius ϵ is the set of all points $\mathbf{x} = (x_1, x_2)$ satisfying $\|\mathbf{x} - \mathbf{a}\| = \sqrt{(x_1 - a_1)^2 + (x_2 - a_2)^2} < \epsilon$. In $\mathcal{R}^3$ an open ball with center at $\mathbf{a} = (a_1, a_2, a_3)$ and positive radius ϵ is the set of all points $\mathbf{x} = (x_1, x_2, x_3)$ satisfying $\|\mathbf{x} - \mathbf{a}\| < \epsilon$. In $\mathcal{R}^n$ an open ball with center at $\mathbf{a} = (a_1, \ldots, a_n)$ and positive radius ϵ is the set of all points $\mathbf{x} = (x_1, \ldots, x_n)$ satisfying $\|\mathbf{x} - \mathbf{a}\| = \sqrt{(x_1 - a_1)^2 + \cdots + (x_n - a_n)^2} < \epsilon$. For general $n \geq 2$ we will mean one of these sets when we talk about an open ball with center at $\mathbf{a}$.

If $f : \mathcal{R}^n \to \mathcal{R}$ then f has a local minimum at a point $\mathbf{a} = (a_1, \ldots, a_n)$ if $f(\mathbf{x}) - f(\mathbf{a}) \geq 0$ for all points $\mathbf{x} = (x_1, \ldots, x_n)$ in some open ball with center at $\mathbf{a}$. The function f has a local maximum at a point $\mathbf{a}$ if $f(\mathbf{x}) - f(\mathbf{a}) \leq 0$ for all points $\mathbf{x}$ in some open ball with center at $\mathbf{a}$. A point $\mathbf{a}$ where f has either a local minimum or maximum will be called an extreme point of f.

Throughout this chapter we will assume that $f : \mathcal{R}^n \to \mathcal{R}$ along with its first and second order partial derivatives are all continuous. In particular it follows that $\mathbf{H}f = [f_{ij}]_{n \times n}$, the Hessian matrix of f of second partial derivatives, is a symmetric matrix.

12.2 Critical Points and Quadratic Approximations at Critical Points

For a continuously differentiable function of one variable to have an extreme point inside an open interval it must have a zero derivative at that point. A similar result holds for $f : \mathcal{R}^n \to \mathcal{R}$.

Theorem 12.1. *Let $f : \mathcal{R}^n \to \mathcal{R}$ be continuous and have continuous partial derivatives in an open ball with center at* $\mathbf{a}$*, If* $\mathbf{a}$ *is an extreme point of f then* $\nabla f(\mathbf{a}) = 0$.

If $f : \mathcal{R}^n \to \mathcal{R}$ and if $\nabla f(\mathbf{a}) = 0$ then $\mathbf{a}$ is called a critical point of f. As in the single variable case every extreme point is a critical point but there also exist saddle points where f has neither a maximum nor a minimum. A function f has a saddle point at a critical point $\mathbf{a} = (a_1, \ldots, a_n)$ if there are pairs of points $\mathbf{x}'$ and $\mathbf{x}''$ inside of every open ball with center at $\mathbf{a}$ for which $f(\mathbf{x}') - f(\mathbf{a}) < 0$ and $f(\mathbf{x}'') - f(\mathbf{a}) > 0$.

Look at the quadratic approximation to f at one of its critical points $\mathbf{a}$. We have $\nabla f(\mathbf{a}) = 0$ so $\mathbf{Q}f(\mathbf{x})$ is given by

$$\mathbf{Q}f(\mathbf{x}) = f(\mathbf{a}) + \frac{1}{2}(\mathbf{x} - \mathbf{a})^T \mathbf{H}f(\mathbf{a})(\mathbf{x} - \mathbf{a}) \quad .$$

Since $f(\mathbf{x}) \approx \mathbf{Q}f(\mathbf{x})$ we have

$$f(\mathbf{x}) - f(\mathbf{a}) \approx \frac{1}{2}(\mathbf{x} - \mathbf{a})^T \mathbf{H}f(\mathbf{a})(\mathbf{x} - \mathbf{a})$$

If $\det \mathbf{H}f(\mathbf{a}) \neq 0$ the sign of $f(\mathbf{x}) - f(\mathbf{a})$ will be the same as the sign of the quadratic form $(\mathbf{x}-\mathbf{a})^T \mathbf{H}f(\mathbf{a})(\mathbf{x}-\mathbf{a})$ near $\mathbf{a}$. If $\det \mathbf{H}f(\mathbf{a}) = 0$ the point $\mathbf{a}$ is called a degenerate critical point of f. A critical point $\mathbf{a}$ with $\det \mathbf{H}f(\mathbf{a}) \neq 0$ is called a non-degenerate or regular critical point of f. The behavior of f at a degenerate critical point is complicated as is illustrated by the following example.

Example 12.1

Let $f(x,y) = x^2 + y^4$, $g(x,y) = x^2 - y^4$, and $h(x,y) = x^2 + y^3$. They all have a critical point at $\mathbf{a} = (0,0)$ since $\nabla f = \langle 2x, 4y^3 \rangle$, $\nabla g = \langle 2x, -4y^3 \rangle$, and $\nabla h = \langle 2x, 3y^2 \rangle$. They all have a zero value at $\mathbf{a} = (0,0)$. Their Hessian matrices are given by

$$\mathbf{H}f = \begin{pmatrix} 2 & 0 \\ 0 & 12y^2 \end{pmatrix} \quad ,$$

$$\mathbf{H}g = \begin{pmatrix} 2 & 0 \\ 0 & -12y^2 \end{pmatrix} \quad ,$$

$$\mathbf{H}h = \begin{pmatrix} 2 & 0 \\ 0 & 6y \end{pmatrix} \quad .$$

At $\mathbf{a} = (0,0)$ the Hessian matrices are all equal to the same singular matrix

$$\mathbf{H}f(0,0) = \mathbf{H}g(0,0) = \mathbf{H}h(0,0) = \begin{pmatrix} 2 & 0 \\ 0 & 0 \end{pmatrix} ,$$

so $\mathbf{a} = (0,0)$ is a degenerate critical point of f, g, and h. Thus, $\mathbf{Q}f(x,y) = \mathbf{Q}g(x,y) = \mathbf{Q}h(x,y) = x^2 \geq 0$, so all three functions have the same quadratic approximation which is always non-negative. But $f(x,y) = x^2 + y^4 \geq 0$; $g(0,y) = -y^4 \leq 0$ and $g(x,0) = x^2 \geq 0$; and $h(0,y) = y^3$ which is positive and negative near $(0,0)$. ∎

Thus we see that in general at a degenerate critical point $\mathbf{a}$,

$$f(\mathbf{x}) - f(\mathbf{a}) \quad \text{and} \quad (\mathbf{x}-\mathbf{a})^T \mathbf{H}f(\mathbf{a})(\mathbf{x}-\mathbf{a})$$

need not necessarily have the same sign. More information about $f(\mathbf{x})$ is needed at such a point.

12.3 Analysis of the Behavior of a Function at Its Critical Points

As is the case for functions of one variable the absolute extreme points for $f : \mathfrak{R}^n \to \mathfrak{R}$ can occur at boundary points or at critical points where $\nabla f(\mathbf{a}) = 0$. We are concerned with the critical points of f in this chapter.

Near a regular critical point $\mathbf{a}$ the signs of

$$f(\mathbf{x}) - f(\mathbf{a}) \quad \text{and} \quad (\mathbf{x}-\mathbf{a})^T \mathbf{H}f(\mathbf{a})(\mathbf{x}-\mathbf{a})$$

are the same, so the eigenvalues of $\mathbf{H}f(\mathbf{a})$ can be used to analyze the behavior of f near a.

Theorem 12.2. *Let $f : \mathfrak{R}^n \to \mathfrak{R}$ be continuous and have continuous first and second partial derivatives in an open ball with center at a critical point $\mathbf{a}$. Then*

(i) f has a local maximum at $\mathbf{a}$ if all the eigenvalues of $\mathbf{H}f(\mathbf{a})$ are negative

(ii) f has a local minimum at $\mathbf{a}$ if all the eigenvalues of $\mathbf{H}f(\mathbf{a})$ are positive

(iii) f has a saddle point at $\mathbf{a}$ if any two eigenvalues of $\mathbf{H}f(\mathbf{a})$ have different signs.

If $\mathbf{a}$ is a degenerate critical point of f and all the eigenvalues of $\mathbf{H}f(\mathbf{a})$ are either non-negative or non-positive more information is needed to analyze the behavior of f near $\mathbf{a}$.

Proof. Since $\mathbf{H}f(\mathbf{a})$ is a symmetric matrix we know by the material in Chapter 8 that we can find a rotation matrix P such that $y = P(\mathbf{x} - \mathbf{a})$ and

$$(\mathbf{x}-\mathbf{a})^T \mathbf{H}f(\mathbf{a})(\mathbf{x}-\mathbf{a}) = \lambda_1 y_1^2 + \cdots + \lambda_n y_n^2$$

where λ_i, $i = 1, \ldots, n$ are the eigenvalues of $\mathbf{H}f(\mathbf{a})$. If all the $\lambda_i < 0$ then $\lambda_1 y_1^2 + \cdots + \lambda_n y_n^2 \leq 0$ and $f(\mathbf{x}) - f(\mathbf{a}) \leq 0$ near $\mathbf{a}$. If all the $\lambda_i > 0$ then $\lambda_1 y_1^2 + \cdots + \lambda_n y_n^2 \geq 0$ and $f(\mathbf{x}) - f(\mathbf{a}) \geq 0$ near $\mathbf{a}$ so f has a local minimum at $\mathbf{a}$. If say $\lambda_1 > 0$ and $\lambda_2 < 0$ choose $\mathbf{x}'$ near $\mathbf{a}$ so that $y_2 = y_3 = \ldots = y_n = 0$ and $f(\mathbf{x}') - f(\mathbf{a}) = \lambda_1 y_1^2 > 0$. Then choose $\mathbf{x}''$ near $\mathbf{a}$ so that $y_1 = y_3 = \cdots = y_n = 0$ and $f(\mathbf{x}'') - f(\mathbf{a}) = \lambda_2 y_2^2 < 0$. Thus, $\mathbf{a}$ is a saddle point for f. Example 12.1 shows why more information is needed at a degenerate critical point. ∎

Example 12.2 For $n = 2$ Theorem 12.2 gives us the following information about a critical point $\mathbf{a} = (a_1, a_2)$ of a function $f = f(x, y)$ in non-matrix terms.

First,

$$\nabla f(\mathbf{a}) = \langle f_1(\mathbf{a}), f_2(\mathbf{a})\rangle = \langle 0, 0\rangle \quad .$$

Then from Chapter 7

$$\mathbf{H}f(\mathbf{a}) = \begin{pmatrix} f_{11}(\mathbf{a}) & f_{12}(\mathbf{a}) \\ f_{12}(\mathbf{a}) & f_{22}(\mathbf{a}) \end{pmatrix}$$

so

$$\operatorname{Tr} \mathbf{H}f(\mathbf{a}) = c_1 = f_{11}(\mathbf{a}) + f_{22}(\mathbf{a}) = \lambda_1 + \lambda_2$$

and

$$\det \mathbf{H}f(\mathbf{a}) = c_2 = f_{11}(\mathbf{a}) f_{22}(\mathbf{a}) - f_{12}^2(\mathbf{a}) = \lambda_1 \lambda_2 \quad .$$

(i) Now, $\mathbf{a}$ is an extreme point of f if λ_1 and λ_2 have the same sign or

$$\lambda_1 \lambda_2 = \det \mathbf{H}f(\mathbf{a}) = f_{11}(\mathbf{a}) f_{22}(\mathbf{a}) - f_{12}^2(\mathbf{a}) > 0 \quad .$$

Then $f_{11}(\mathbf{a}) f_{22}(\mathbf{a}) = \lambda_1 \lambda_2 + f_{12}^2(\mathbf{a}) > 0$ so $f_{11}(\mathbf{a})$ and $f_{22}(\mathbf{a})$ both have the same sign as λ_1 and λ_2. Thus,

1. if $f_{11}(\mathbf{a}) < 0$ then f has a local maximum at $\mathbf{a}$
2. if $f_{11}(\mathbf{a}) > 0$ then f has a local minimum at $\mathbf{a}$.

(ii) If $\lambda_1\lambda_2 = \det \mathbf{H}f(\mathbf{a}) < 0$ then λ_1 and λ_2 have opposite signs so $\mathbf{a}$ is a saddle point of f.

(iii) If $\lambda_1\lambda_2 = \det \mathbf{H}f(\mathbf{a}) = 0$ then at least one eigenvalue is zero and more information about $f(x,y)$ is needed. ∎

For $n \geq 3$ there are similar complicated and obsolete criteria which could be derived from the eigenvalues of $\mathbf{H}f(\mathbf{a})$. There is no reason to do this.

The following examples will illustrate the use of Theorem 12.2.

Example 12.3

Find the critical points of

$$f(x,y) = -21 + 24y - 4x^2 - 4xy - 7y^2$$

and use the eigenvalues of $\mathbf{H}f$ to identify the relative maxima and minima of f.

Solution.

From $\nabla f = \langle -8x - 4y, -4x - 14y + 24\rangle = \langle 0, 0\rangle$ it follows that

$$\begin{aligned} -8x - 4y &= 0 \\ -4x - 14y &= -24 \end{aligned}$$

or

$$\begin{aligned} 2x + y &= 0 \\ 2x + 7y &= 12 \end{aligned}$$

which has one solution $(x,y) = (-1,2)$. Since

$$\mathbf{H}f = \begin{pmatrix} -8 & -4 \\ -4 & -14 \end{pmatrix}$$

is a constant, it equals $A = \mathbf{H}f(-1,2)$. Now we need to find the eigenvalues of A. From $c_1 = \operatorname{Tr} A = -8 - 14 = -22$ and $c_2 = \det A = 112 - 16 = 96$ we have $f_A(\lambda) = \lambda^2 - c_1\lambda + c_2 = \lambda^2 + 22\lambda + 96 = (\lambda + 6)(\lambda + 16) = 0$. Thus, $\lambda_1 = -6$ and $\lambda_2 = -16$ are the eigenvalues of A. Since they are both negative, f has a relative maximum at $(-1,2)$. ∎

Example 12.4

Find the critical points of

$$f(x,y) = -4x^2 + 6xy + 4y^2$$

and use the eigenvalues of $\mathbf{H}f$ to analyze the critical points.

Solution.

From $\nabla f = \langle -8x + 6y, 6x + 8y \rangle = \langle 0, 0 \rangle = \mathbf{0}$ we obtain the system

$$\begin{aligned} -8x + 6y &= 0 \\ 6x + 8y &= 0 \end{aligned} .$$

The augmented matrix of the system is

$$\left(\begin{array}{cc|c} -8 & 6 & 0 \\ 6 & 8 & 0 \end{array}\right) \sim \left(\begin{array}{cc|c} -4 & 3 & 0 \\ 3 & 4 & 0 \end{array}\right) \sim \left(\begin{array}{cc|c} 1 & -1 & 0 \\ 0 & 25 & 0 \end{array}\right) \sim \left(\begin{array}{cc|c} 1 & 0 & 0 \\ 0 & 1 & 0 \end{array}\right) .$$

Thus $(x, y) = (0, 0)$ is the only critical point of f. The matrix $\mathbf{H}f$ is a constant and hence

$$A = \mathbf{H}f(0,0) = \begin{pmatrix} -8 & 6 \\ 6 & 8 \end{pmatrix} .$$

Use $c_1 = \operatorname{Tr} A = -8 + 8 = 0$ and $c_2 = \det A = -64 - 36 = -100$ to find $f_A(\lambda) = \lambda^2 - 100 = (\lambda - 10)(\lambda + 10) = 0$. The eigenvalues are $\lambda_1 = 10$ and $\lambda_2 = -10$. Since they have opposite signs $(0, 0)$ is a saddle point of f. ∎

Example 12.5

Find the critical points of

$$f(x,y,z) = 12 + 16x - 10y + 6z + 8x^2 - 10xy + 6xz + 13y^2 - 4yz + 5z^2$$

and use the eigenvalues of $\mathbf{H}f$ to analyse the critical points.

Solution.

From $\nabla f = \langle 16 + 16x - 10y + 6z, -10 - 10x + 26y - 4z, 6 + 6x - 4y + 10z \rangle = \mathbf{0}$ we obtain the system

$$\begin{aligned} 3x - 2y + 5z &= -3 \\ 5x - 13y + 2z &= -5 \\ 8x - 5y + 3z &= -8 \ . \end{aligned}$$

The augmented matrix of the system is

$$\left(\begin{array}{rrr|r} 3 & -2 & 5 & -3 \\ 5 & -13 & 2 & -5 \\ 8 & -5 & 3 & -8 \end{array}\right)$$

which can be row reduced to

$$\left(\begin{array}{rrr|r} 1 & -9 & 8 & -1 \\ 0 & 10 & -4 & 0 \\ 0 & 32 & -38 & 0 \end{array}\right) \sim \left(\begin{array}{rrr|r} 1 & 0 & 0 & -1 \\ 0 & 1 & 0 & 0 \\ 0 & 0 & 1 & 0 \end{array}\right) \ .$$

Thus, $(x, y, z) = (-1, 0, 0)$ is the solution to the system and the only critical point of f. The Hessian matrix $\mathbf{H}f$ is a constant and hence

$$A = \mathbf{H}f(-1,0,0) = \begin{pmatrix} 16 & -10 & 6 \\ -10 & 26 & -4 \\ 6 & -4 & 10 \end{pmatrix} \ .$$

use the principal minors of A to find c_1, c_2, and c_3 from the following:

$$c_1 = 16 + 26 + 10 \quad = \quad 52$$

$$c_2 = \begin{vmatrix} 16 & -10 \\ -10 & 26 \end{vmatrix} + \begin{vmatrix} 16 & 5 \\ 6 & 10 \end{vmatrix} + \begin{vmatrix} 26 & -4 \\ -4 & 10 \end{vmatrix} = 316 + 124 + 244 = 684$$

$$c_3 = \begin{vmatrix} 16 & -10 & 6 \\ -10 & 26 & -4 \\ 6 & -4 & 10 \end{vmatrix} = 2448 \ .$$

Thus, $f_A(\lambda) = \lambda^3 - c_1\lambda^2 + c_2\lambda - c_3 = \lambda^3 - 52\lambda^2 + 684\lambda - 2448 = 0$ is the characteristic equation of A.

The roots of $f_A(\lambda) = 0$ are $\lambda_1 = 34$, $\lambda_2 = 12$, and $\lambda_3 = 6$. Since the eigenvalues of $A = \mathbf{H}f(-1,0,0)$ are all positive it follows that f has a relative minimum at $(-1,0,0)$.

Example 12.6

Show that the critical points of

$$f(x,y,z) = x^2 + y^2 + 2z^2 - 2xyz$$

are $(0,0,0)$, $(\sqrt{2},\sqrt{2},1)$, $(-\sqrt{2},-\sqrt{2},1)$, $(\sqrt{2},-\sqrt{2},-1)$, and $(-\sqrt{2},\sqrt{2},-1)$. Then use the eigenvalues of $\mathbf{H}f$ to analyze the critical points $(0,0,0)$, $(\sqrt{2},\sqrt{2},1)$, and $(\sqrt{2},-\sqrt{2},-1)$.

Solution.

From

$$\nabla f = \langle 2x - 2yz, 2y - 2xz, 4z - 2xy \rangle = \langle 0,0,0 \rangle = \mathbf{0}$$

we have

$$\begin{aligned} 2x - 2yz &= 0 \\ 2y - 2xz &= 0 \\ 4z - 2xy &= 0 \end{aligned}$$

or

$$x = yz, \qquad y = xz, \quad \text{and} \quad 2z = xy \quad .$$

It follows that $x^2 = y^2 = 2z^2 = xyz$. If $xyz = 0$ then $x = y = z = 0$. If $xyz \neq 0$ then

$$z = \frac{x}{y} = \frac{y}{x} = \frac{xy}{2} \quad .$$

It follows that $x^2 = y^2 = 2$ and $z^2 = 1$. The following points satisfy these conditions: $(\sqrt{2},\sqrt{2},1)$, $(-\sqrt{2},-\sqrt{2},1)$, $(\sqrt{2},-\sqrt{2},-1)$, and $(-\sqrt{2},\sqrt{2},-1)$. Thus we have found the five critical points of f.

The matrix $\mathbf{H}f$ is given by

$$\mathbf{H}f = \begin{pmatrix} 2 & -2z & -2y \\ -2z & 2 & -2x \\ -2y & -2x & 4 \end{pmatrix} \quad .$$

Since

$$\mathbf{H}f(0,0,0) = \begin{pmatrix} 2 & 0 & 0 \\ 0 & 2 & 0 \\ 0 & 0 & 4 \end{pmatrix}$$

has positive eigenvalues 2, 2, and 4 it follows that f has a relative minimum at $(0,0,0)$.

For $(\sqrt{2},\sqrt{2},1)$ we have

$$A = \mathbf{H}f(\sqrt{2},\sqrt{2},1) = \begin{pmatrix} 2 & -2 & -2\sqrt{2} \\ -2 & 2 & -2\sqrt{2} \\ -2\sqrt{2} & -2\sqrt{2} & 4 \end{pmatrix} .$$

Using the principal minors of A we obtain c_1, c_2, and c_3 from the following:

$$c_1 = 2 + 2 + 4 = 8$$

$$c_2 = \begin{vmatrix} 2 & -2 \\ -2 & 2 \end{vmatrix} + \begin{vmatrix} 2 & -2\sqrt{2} \\ -2\sqrt{2} & 4 \end{vmatrix} + \begin{vmatrix} 2 & -2\sqrt{2} \\ -2\sqrt{2} & 4 \end{vmatrix} = 0 + 0 + 0 = 0$$

$$c_3 = \begin{vmatrix} 2 & -2 & -2\sqrt{2} \\ -2 & 2 & -2\sqrt{2} \\ -2\sqrt{2} & -2\sqrt{2} & 4 \end{vmatrix} = 8 \begin{vmatrix} 1 & -1 & -\sqrt{2} \\ -1 & 1 & -\sqrt{2} \\ -\sqrt{2} & -\sqrt{2} & 2 \end{vmatrix}$$

$$= 8 \begin{vmatrix} 1 & -1 & -\sqrt{2} \\ 0 & 0 & -2\sqrt{2} \\ 0 & -2\sqrt{2} & 0 \end{vmatrix} = -64 \quad .$$

Thus, $f_A(\lambda) = \lambda^3 - c_2\lambda^2 + c_2\lambda - c_3 = \lambda^3 - 8\lambda^2 + 64 = 0$ is the characteristic equation of A. The eigenvalues are $\lambda_1 = 4$, $\lambda_2 = 2 + 2\sqrt{5}$, and $\lambda_3 = 2 - 2\sqrt{5}$. Since λ_1 and λ_2 are positive and λ_3 is negative it follows that f has a saddle point at $(\sqrt{2},\sqrt{2},1)$.

For $(\sqrt{2},-\sqrt{2},-1)$ we have

$$B = \mathbf{H}f(\sqrt{2},-\sqrt{2},-1) = \begin{pmatrix} 2 & 2 & 2\sqrt{2} \\ 2 & 2 & -2\sqrt{2} \\ 2\sqrt{2} & -2\sqrt{2} & 4 \end{pmatrix} .$$

Using the principal minors of B we obtain c_1, c_2, and c_3 from the following:

$$c_1 = 2 + 2 + 4 = 8$$

$$c_2 = \begin{vmatrix} 2 & 2 \\ 2 & 2 \end{vmatrix} + \begin{vmatrix} 2 & 2\sqrt{2} \\ 2\sqrt{2} & 4 \end{vmatrix} + \begin{vmatrix} 2 & -2\sqrt{2} \\ -2\sqrt{2} & 4 \end{vmatrix} = 0 + 0 + 0 = 0$$

$$c_3 = \begin{vmatrix} 2 & 2 & 2\sqrt{2} \\ 2 & 2 & -2\sqrt{2} \\ 2\sqrt{2} & -2\sqrt{2} & 4 \end{vmatrix} = 8 \begin{vmatrix} 1 & 1 & \sqrt{2} \\ 1 & 1 & -\sqrt{2} \\ \sqrt{2} & -\sqrt{2} & 2 \end{vmatrix}$$

$$= 8 \begin{vmatrix} 1 & 1 & \sqrt{2} \\ 0 & 0 & -2\sqrt{2} \\ 0 & -2\sqrt{2} & 0 \end{vmatrix} = -64 \quad .$$

Thus, $f_B(\lambda) = \lambda^3 - 8\lambda^2 + 64 = 0$ is the characteristic equation of B. It is the same as the one for A so the same conclusion holds, namely, f has a saddle point at $(\sqrt{2}, -\sqrt{2}, -1)$. ∎

As can be seen from these examples the critical points of a function f of several variables can be analyzed by finding the eigenvalues of $\mathbf{H}f$ at the critical points. There are many good software packages for the computation of the eigenvalues of symmetric matrices, so this is a practical way to analyze critical points.

There are other techniques for analyzing quadratic forms, not requiring the eigenvalues, which are covered in more advanced linear algebra texts.

12.4 Exercises

1. Find the critical points of f and use the eigenvalues of $\mathbf{H}f$ to analyze them if

a) $f(x, y) = 42 + 16x - 16y + 7x^2 + 4xy + 4y^2$

b) $f(x, y) = -45 + 14x - 5x^2 - 12xy - 10y^2$

c) $f(x, y) = 4 - x - 7y - x^2 - 5y^2 - 3xy$

d) $f(x, y) = \frac{1}{3}x^3 - x^2 + \frac{1}{4}y^4 - 2y^2$

2. Find the critical points of f and use the eigenvalues of $\mathbf{H}f$ to analyze them if

a) $f(x, y, z) = 3x^2 + 4xy + 6y^2 - 2x + 8y + 7 + z^2 + 2z$

b) $f(x,y,z) = x^2 + y^2 + z^2 - xy + 3x - 2z$

c) $f(x,y,z) = x^2 + y^2 + z^2 + xy - y - 14z$

d) $f(x,y,z) = x^2 + y^2 + z^2 - xy + x - z$

e) $f(x,y,z) = -\frac{3}{2}x^2 + 3xy - \frac{11}{2}y^2 + 3x + 5y - z^2 - 4z$

f) $f(x,y,z) = -5x^2 + 4xy - 2y^2 - 2x + 8y + z^3 - 3z + 6$

g) $f(x,y,z) = 2x^3 + 3x^2 - 12x + y^3 - 3y^2 - 9y + z^2 - 4z$

h) $f(x,y,z) = x^2 + xy + y^2 + x - 4y + 5 + 4z^2 - 8z$

i) $f(x,y,z) = \frac{1}{3}x^3 - \frac{3}{2}x^2 + \frac{1}{4}y^4 - \frac{9}{2}y^2 + z^2 + 4z$

3. A function $f : \mathcal{R}^3 \to \mathcal{R}$ has a critical point at $\mathbf{a}$ where $\nabla f(\mathbf{a}) = 0$. If

$$\mathbf{H}f(\mathbf{a}) = \begin{pmatrix} 1+c & 0 & 0 \\ 0 & c & 0 \\ 0 & 0 & c-1 \end{pmatrix}$$

give conditions on c for which

a) f has a relative maximum at $\mathbf{a}$

b) f has a relative minimum at $\mathbf{a}$

c) f has a saddle point at $\mathbf{a}$

d) $\mathbf{a}$ is a degenerate critical point of f.

4. The eigenvectors of $\mathbf{H}f$ are given in this problem. Use matrix multiplication to determine the eigenvalues of $\mathbf{H}f$ at the critical points of f. If

$$f(x,y,z) = 3x + 2y + 4z - 5y^2 - x^3 + 4yz - 2z^2$$

the eigenvectors of $\mathbf{H}f$ at any point are given by

$$\{\mathbf{p}_1 = \begin{pmatrix} 1 \\ 0 \\ 0 \end{pmatrix}, \quad \mathbf{p}_2 = \begin{pmatrix} 0 \\ 1 \\ 2 \end{pmatrix}, \quad \mathbf{p}_3 = \begin{pmatrix} 0 \\ 2 \\ -1 \end{pmatrix}\} .$$

Find the critical points of f and use the eigenvalues of $\mathbf{H}f$ to analyze them.

5. If

$$f(x,y,z) = 2x - 6y - z + \frac{1}{2}x^2 - \frac{1}{3}x^3 + 3y^2 + 6yz + \frac{11}{2}z^2$$

the eigenvectors of $\mathbf{H}f$ at any point are given by

$$\{\mathbf{p}_1 = \begin{pmatrix} 1 \\ 0 \\ 0 \end{pmatrix}, \quad \mathbf{p}_2 = \begin{pmatrix} 0 \\ 2 \\ 3 \end{pmatrix}, \quad \mathbf{p}_3 = \begin{pmatrix} 0 \\ 3 \\ -2 \end{pmatrix}\} .$$

Repeat Problem 4 for this function.

6. A function $f : \mathcal{R}^3 \to \mathcal{R}$ has critical points at $\mathbf{a}$ and $\mathbf{b}$. If $A = \mathbf{H}f(\mathbf{a})$ and $B = \mathbf{H}f(\mathbf{b})$ have eigenvectors $\{\mathbf{p}_1, \mathbf{p}_2, \mathbf{p}_3\}$, use matrix multiplication to find the eigenvalues and then use them to analyze the critical points if

a) $A = \begin{pmatrix} 4 & -1 & 3 \\ -1 & 8 & -1 \\ 3 & -1 & 4 \end{pmatrix}, \quad B = \begin{pmatrix} -6 & -1 & 3 \\ -1 & -2 & -1 \\ 3 & -1 & -6 \end{pmatrix},$

$$\mathbf{p}_1 = \begin{pmatrix} 1 \\ 0 \\ -1 \end{pmatrix}, \quad \mathbf{p}_2 = \begin{pmatrix} 1 \\ 1 \\ 1 \end{pmatrix}, \quad \mathbf{p}_3 = \begin{pmatrix} 1 \\ -2 \\ 1 \end{pmatrix}$$

b) $A = \begin{pmatrix} -2 & 1 & -3 \\ 1 & -6 & 1 \\ -3 & 1 & -2 \end{pmatrix}, \quad B = \begin{pmatrix} 8 & 1 & -3 \\ 1 & 4 & 1 \\ -3 & 1 & 8 \end{pmatrix},$

$$\mathbf{p}_1 = \begin{pmatrix} 1 \\ 0 \\ -1 \end{pmatrix}, \quad \mathbf{p}_2 = \begin{pmatrix} 1 \\ 1 \\ 1 \end{pmatrix}, \quad \mathbf{p}_3 = \begin{pmatrix} 1 \\ -2 \\ 1 \end{pmatrix} .$$

7. For each of the following functions show that $(0,0,0)$ is a critical point and find all the other critical points. Then use the eigenvalues of $\mathbf{H}f$ at $(0,0,0)$ and at the critical point $\mathbf{a} = (a_1, a_2, a_3)$ with positive coordinates to analyze them if

a) $f(x,y,z) = x^2 + y^2 + z^2 - 2xyz$

b) $f(x,y,z) = -x^2 - y^2 - z^2 + 2xyz$

c) $f(x,y,z) = x^2 + y^2 + z^2 - xyz$

d) $f(x,y,z) = -x^2 - 3y^2 - 3z^2 + 2\sqrt{3}\,xyz$

e) $f(x,y,z) = x^2 + 3y^2 + 3z^2 - \sqrt{3}\,xyz$
f) $f(x,y,z) = -x^2 - 6y^2 - 6z^2 + \sqrt{6}\,xyz$
g) $f(x,y,z) = x^2 + 6y^2 + 6z^2 - 2\sqrt{6}\,xyz$
h) $f(x,y,z) = x^2 + 3y^2 + 3z^2 - 2\sqrt{3}\,xyz$.

8. For each of the following functions show that $(0,0,0)$ is a critical point and find all the other critical points. Then use the eigenvalues of $\mathbf{H}f$ at $(0,0,0)$ and at the critical point $\mathbf{a} = (a_1, a_2, a_3)$ with all negative coordinates to analyze them if

a) $f(x,y,z) = x^2 + y^2 + z^2 + 2xyz$
b) $f(x,y,z) = -x^2 - y^2 - z^2 - 2xyz$
c) $f(x,y,z) = x^2 + y^2 + z^2 + xyz$
d) $f(x,y,z) = x^2 + 3y^2 + 3z^2 + 2\sqrt{3}\,xyz$
e) $f(x,y,z) = -x^2 - 3y^2 - 3z^2 - \sqrt{3}\,xyz$
f) $f(x,y,z) = x^2 + 6y^2 + 6z^2 + \sqrt{6}\,xyz$
g) $f(x,y,z) = -x^2 - 6y^2 - 6z^2 - 2\sqrt{6}\,xyz$
h) $f(x,y,z) = x^2 + 3y^2 + 3z^2 + 2\sqrt{3}\,xyz$.

9. For each of the following functions find all the critical points and use the values of $\mathbf{H}f$ to analyze them if

a) $f(x,y,z) = x^2 + y^2 - z^2 - 2xyz$
b) $f(x,y,z) = x^2 + y^2 - z^2 + xyz$
c) $f(x,y,z) = x^2 - 3y^2 - 3z^2 + 2\sqrt{3}\,xyz$
d) $f(x,y,z) = x^2 - 6y^2 - 6z^2 - \sqrt{6}\,xyz$.

10. Show that $(0,0,0)$ is a degenerate critical point of $f(x,y,z) = x^2 + y^2 + z^3$, $g(x,y,z) = x^2 + y^2 + z^4$, and $h(x,y,z) = x^2 + y^2 - z^4$, and that they all have the same non-negative quadratic approximation at $(0,0,0)$. Then show that f and h have saddle points at $(0,0,0)$ and that g has a minimum at $(0,0,0)$.

13 Lagrange Multiplier Problems with Quadratic Functions and Quadratic Restraints

Outline of Key Ideas

Functions of Two variables.

Functions of Three variables.

Functions of n variables.

Lagrange Multiplier Problems with Quadratic Functions and Quadratic Restraints

13.1 Functions of Two Variables

If $f(x,y)$ is a function of two variables and if $g(x,y) = c$ is a constraint then the critical points of $f(x,y)$ subject to this constraint are the points (x,y) which satisfy the Lagrange multiplier equations

$$\nabla f = \lambda \nabla g \quad \text{and} \quad g(x,y) = c \quad .$$

The following example shows that if A is a 2×2 symmetric matrix then its eigenvalue problem is a special case of a Lagrange multiplier problem with a quadratic function $f(x,y)$ and a quadratic restraint $g(x,y) = x^2 + y^2 = c > 0$.

Example 13.1 Let

$$A = \begin{pmatrix} a_{11} & a_{12} \\ a_{12} & a_{22} \end{pmatrix}$$

be a symmetric matrix and let

$$f(x_1, x_2) = \mathbf{x}^T A \mathbf{x} = a_{11}x_1^2 + 2a_{12}x_1x_2 + a_{22}x_2^2$$

subject to the constraint

$$f(x_1, x_2) = \mathbf{x}^T \mathbf{x} = x_1^2 + x_2^2 = c > 0 \quad .$$

Show that the critical points of f with this constraint are given by the eigenvectors $\mathbf{x} = [x_1, x_2]^T$ of length $\sqrt{c}$ where the Lagrange multiplier λ is the corresponding eigenvalue and $f(x_1, x_2) = \lambda c$.

Solution.

For $f(x_1, x_2) = a_{11}x_1^2 + 2a_{12}x_1x_2 + a_{22}x_2^2$ with the constraint $g(x_1, x_2) = x_1^2 + x_2^2 = c > 0$ the corresponding Lagrange multiplier equations are

$$\nabla f = \langle 2a_{11}x_1 + 2a_{12}x_2, 2a_{12}x_1 + 2a_{22}x_2 \rangle = 2\lambda \langle x_1, x_2 \rangle = \lambda \nabla g \qquad (13.1)$$

and

$$g(x_1, x_2) = x_1^2 + x_2^2 = c \quad . \tag{13.2}$$

After simplifying (13.1) we obtain

$$\langle a_{11}x_1 + a_{12}x_2, a_{12}x_1 + a_{22}x_2 \rangle = \lambda \langle x_1, x_2 \rangle \tag{13.3}$$

which is equivalent to the system

$$(a_{11} - \lambda)x_1 + a_{12}x_2 = 0 \tag{13.4}$$

$$a_{12}x_1 + (a_{22} - \lambda)x_2 = 0 \tag{13.5}$$

where $g(x_1, x_2) = x_1^2 + x_2^2 = c$. In matrix form this is

$$\begin{pmatrix} a_{11} - \lambda & a_{12} \\ a_{12} & a_{22} - \lambda \end{pmatrix} \begin{pmatrix} x_1 \\ x_2 \end{pmatrix} = \begin{pmatrix} 0 \\ 0 \end{pmatrix} \tag{13.6}$$

or

$$(A - \lambda I)\mathbf{x} = 0 \tag{13.7}$$

where $\mathbf{x} = [x_1, x_2]^T$ is a vector of length $\sqrt{c} \neq 0$ so $\mathbf{x} \neq \mathbf{0}$. This is just the eigenvalue problem from Chapter 7. If $\mathbf{x}$ is a solution of (13.7) then $g(\mathbf{x}) = \mathbf{x}^T\mathbf{x} = c$ and $A\mathbf{x} = \lambda\mathbf{x}$. Since $f(\mathbf{x}) = \mathbf{x}^T\mathbf{x}$ it follows that

$$f(\mathbf{x}) = \mathbf{x}^T(A\mathbf{x}) = \mathbf{x}^T(\lambda\mathbf{x}) = \lambda\mathbf{x}^T\mathbf{x} = \lambda g(\mathbf{x}) = \lambda c \quad . \quad \blacksquare$$

The following is a specific example to illustrate these ideas.

Example 13.2 Use Lagrange multipliers to find the critical points of $f(x, y) = x^2 + 12xy + 6y^2$ subject to the constraint $g(x, y) = x^2 + y^2 = 13$.

Solution.

The Lagrange multiplier equations are

$$\nabla f = \langle 2x + 12y, 12x + 12y \rangle = \lambda \langle 2x, 2y \rangle = \lambda \nabla g$$

and

$$g(x, y) = x^2 + y^2 = 13 \quad .$$

After simplification we obtain

$$\langle x + 6y, 6x + 6y \rangle = \lambda \langle x, y \rangle$$

which is equivalent to the system

$$\begin{aligned} (1-\lambda)x + 6y &= 0 \\ 6x + (6-\lambda)y &= 0 \end{aligned}$$

where $x^2 + y^2 = 13$. In matrix form the system is given by $(A - \lambda I)\mathbf{x} = 0$ or

$$\begin{pmatrix} 1-\lambda & 6 \\ 6 & 6-\lambda \end{pmatrix} \begin{pmatrix} x \\ y \end{pmatrix} = \begin{pmatrix} 0 \\ 0 \end{pmatrix} \tag{13.8}$$

where $\mathbf{x} = [x, y]^T$ has length $\sqrt{13}$ and A is the matrix

$$A = \begin{pmatrix} 1 & 6 \\ 6 & 6 \end{pmatrix} .$$

This is an eigenvalue problem for A where the Lagrange multiplier λ is an eigenvalue and $\mathbf{x}$ is an eigenvector of length $\sqrt{13}$. In order to solve (13.8) find the characteristic equation of A

$$f_A(\lambda) = \lambda^2 - 7\lambda - 30 = 0 \quad . \tag{13.9}$$

Since

$$\lambda^2 - 7\lambda - 30 = (\lambda + 3)(\lambda - 10)$$

the eigenvalues of A are $\lambda_1 = 10$ and $\lambda_2 = -3$.

Corresponding to $\lambda_1 = 10$ we have

$$A - 10I = \begin{pmatrix} -9 & 6 \\ 6 & -4 \end{pmatrix} \sim \begin{pmatrix} 3 & -2 \\ 0 & 0 \end{pmatrix} \sim \begin{pmatrix} 1 & \dfrac{-2}{3} \\ 0 & 0 \end{pmatrix} .$$

Thus, $x - \dfrac{2}{3}y = 0$. Set $y = 3a$ $(a \neq 0)$ to obtain $\langle x, y \rangle^T = \langle 2a, 3a \rangle^T$. Put this into $g(x, y) = x^2 + y^2 = 13$ to obtain $4a^2 + 9a^2 = 13a^2 = 13$ or $a = \pm 1$. Then $\langle x, y \rangle^T = \langle \pm 2, \pm 3 \rangle^T$ are the eigenvectors of length $\sqrt{13}$ corresponding to $\lambda_1 = 10$ and $f(\pm 2, \pm 3) = 10g(\pm 2, \pm 3) = 130$.

Corresponding to $\lambda_2 = -3$ we have

$$A + 3I = \begin{pmatrix} 4 & 6 \\ 6 & 9 \end{pmatrix} \sim \begin{pmatrix} 2 & 3 \\ 0 & 0 \end{pmatrix} \sim \begin{pmatrix} 1 & \frac{3}{2} \\ 0 & 0 \end{pmatrix} .$$

Thus, $x + \frac{3}{2}y = 0$. Set $y = 2b$ $(b \neq 0)$ to obtain $\langle x, y\rangle^T = \langle -3b, 2b\rangle^T$. Put this into $g(x,y) = x^2 + y^2 = 13$ to get $9b^2 + 4b^2 = 13b^2 = 13$ or $b = \pm 1$. Then $\langle x, y\rangle^T = \langle \mp 3, \pm 2\rangle^T$ are the eigenvectors of length $\sqrt{13}$ corresponding to $\lambda_2 = -3$ and $f(\mp 3, \pm 2) = -3g(\mp 3, \pm 2) = -39$.

Thus, $-39 \leq f(x,y) \leq 130$ when $x^2 + y^2 = 13$. The maximum value of f occurs at $(\pm 2, \pm 3)$ and the minimum value of f occurs at $(x,y) = (\mp 3, \pm 2)$. ∎

In Examples (13.1) and (13.2) we saw that the critical points (x_1, x_2) satisfy

$$f(x_1, x_2) = \mathbf{x}^T A\mathbf{x} = \lambda g(x_1, x_2) = \lambda \mathbf{x}^T\mathbf{x} = \lambda c$$

where $\mathbf{x} = \langle x_1, x_2\rangle^T$ was an eigenvector of length $\sqrt{c}$ corresponding to an eigenvalue λ of the matrix A. Both these examples are special cases of the following general problem.

Example 13.3

If A and B are 2×2 symmetric matrices find the critical points of the quadratic function $f(x_1, x_2) = \mathbf{x}^T A\mathbf{x}$ subject to the quadratic restraint $g(x_1, x_2) = \mathbf{x}^T B\mathbf{x} = c \neq 0$. Thus, if f is given by

$$f(x_1, x_2) = a_{11}x_1^2 + 2a_{12}x_1x_2 + a_{22}x_2^2$$

and subject to the constraint

$$g(x_1, x_2) = b_{11}x_1^2 + 2b_{12}x_1x_2 + b_{22}x_2^2 = c \neq 0$$

we want to determine the critical points (x_1, x_2).

Solution.

The corresponding Lagrange multiplier equations are

$$\begin{aligned} \nabla f &= \langle 2a_{11}x_1 + 2a_{12}x_2, 2a_{12}x_1 + 2a_{22}x_2\rangle \\ &= \lambda\langle 2b_{11}x_1 + 2b_{12}x_2, 2b_{12}x_1 + 2b_{22}x_2\rangle \quad = \quad \lambda\nabla g \end{aligned}$$

and

$$g(x_1,x_2) = b_{11}x_1^2 + 2b_{12}x_1x_2 + b_{22}x_2^2 = c \quad .$$

After simplifying these equations and writing them in matrix form we obtain

$$\begin{pmatrix} a_{11} & a_{12} \\ a_{12} & a_{22} \end{pmatrix} \begin{pmatrix} x_1 \\ x_2 \end{pmatrix} = \lambda \begin{pmatrix} b_{11} & b_{12} \\ b_{12} & b_{22} \end{pmatrix} \begin{pmatrix} x_1 \\ x_2 \end{pmatrix}$$

or

$$\begin{pmatrix} a_{11} - \lambda b_{11} & a_{12} - \lambda b_{12} \\ a_{12} - \lambda b_{12} & a_{22} - \lambda b_{22} \end{pmatrix} \begin{pmatrix} x_1 \\ x_2 \end{pmatrix} = \begin{pmatrix} 0 \\ 0 \end{pmatrix} \tag{13.10}$$

where $g(x_1,x_2) = b_{11}x_1^2 + 2b_{12}x_1x_2 + b_{22}x_2^2 = c \neq 0$.

In terms of A and B we have

$$(A - \lambda B)\mathbf{x} = \mathbf{0} \tag{13.11}$$

where $g(x_1,x_2) = \mathbf{x}^T B\mathbf{x} \neq 0$. The vector $\mathbf{x} \neq 0$, since $g(x_1,x_2) \neq 0$, so the coefficient matrix $A - \lambda B$ in (13.11) must be singular. Thus, $\det(A - \lambda B)$ must equal zero. The Lagrange multipliers λ are the roots of the polynomial equation $\det(A - \lambda B) = 0$. Corresponding to each root λ solve the system (13.11) for vectors $\mathbf{x}$'s satisfying $\mathbf{x}^T B\mathbf{x} = c$. If $(A - \lambda B)\mathbf{x} = \mathbf{0}$ and $\mathbf{x}^T B\mathbf{x} = c$ then

$$f(x_1,x_2) = \mathbf{x}^T A\mathbf{x} = \mathbf{x}^T(\lambda B\mathbf{x}) = \lambda \mathbf{x}^T B\mathbf{x} = \lambda g(x_1,x_2) = \lambda c. \quad \blacksquare$$

The following is a specific example to illustrate these ideas

Example 13.4 Use Lagrange multipliers to find the critical points of $f(x,y) = x^2 + 4xy + 2y^2$ subject to the constraint $g(x,y) = 3x^2 + 2xy + y^2 = 2$.

Solution.

The corresponding Lagrange multiplier equations are

$$\nabla f = \langle 2x + 4y, 4x + 4y \rangle = \lambda \langle 6x + 2y, 2x + 2y \rangle = \lambda \nabla g$$

and $g(x,y) = 3x^2 + 2xy + y^2 = 2$. After simplifying these equations we obtain

$$\langle x + 2y, 2x + 2y \rangle = \lambda \langle 3x + y, x + y \rangle$$

or

$$\begin{pmatrix} 1 & 2 \\ 2 & 2 \end{pmatrix} \begin{pmatrix} x \\ y \end{pmatrix} = \lambda \begin{pmatrix} 3 & 1 \\ 1 & 1 \end{pmatrix} \begin{pmatrix} x \\ y \end{pmatrix}$$

in matrix form where $g(x, y) = 3x^2 + 2xy + y^2 = 2$. Further simplification gives us

$$\begin{pmatrix} 1-3\lambda & 2-\lambda \\ 2-\lambda & 2-\lambda \end{pmatrix} \begin{pmatrix} x \\ y \end{pmatrix} = \begin{pmatrix} 0 \\ 0 \end{pmatrix} \tag{13.12}$$

where $3x^2 + 2xy + y^2 = 2$. Since $(x, y) \neq (0, 0)$ the determinant of the coefficient matrix must be set equal to zero to obtain

$$(1 - 3\lambda)(2 - \lambda) - (2 - \lambda)(2 - \lambda) = 0$$

or

$$(1 - 3\lambda - 2 + \lambda)(2 - \lambda) = (-1 - 2\lambda)(2 - \lambda) = (\lambda - 2)(2\lambda + 1) = 0 \quad .$$

Thus, the Lagrange multipliers are $\lambda_1 = 2$ and $\lambda_2 = -\frac{1}{2}$.

Corresponding to $\lambda_1 = 2$ we have

$$\begin{pmatrix} 1-6 & 2-2 \\ 2-2 & 2-2 \end{pmatrix} = \begin{pmatrix} -5 & 0 \\ 0 & 0 \end{pmatrix} \sim \begin{pmatrix} 1 & 0 \\ 0 & 0 \end{pmatrix} \quad .$$

Thus, $x = 0$ and $y = a \neq 0$. Put $\langle 0, a\rangle^T$ into $g(0, a) = 0 + 0 + a^2 = a^2 = 2$ to obtain $a = \pm\sqrt{2}$. Therefore, the corresponding critical points of f are $(0, \pm\sqrt{2})$.

Corresponding to $\lambda_2 = -\frac{1}{2}$ we have

$$\begin{pmatrix} 1+\frac{3}{2} & 2+\frac{1}{2} \\ 2+\frac{1}{2} & 2+\frac{1}{2} \end{pmatrix} = \frac{1}{2} \begin{pmatrix} 5 & 5 \\ 5 & 5 \end{pmatrix} \sim \begin{pmatrix} 1 & 1 \\ 0 & 0 \end{pmatrix} \quad .$$

Thus, $x + y = 0$, set $y = b \neq 0$ then $x = -b$ and put $\langle -b, b\rangle$ into $g(-b, b) = 3b^2 - 2b^2 + b^2 = 2b^2 = 2$ to obtain $b = \pm 1$. Therefore, the corresponding critical points of f are $(\pm 1, \mp 1)$.

With

$$A = \begin{pmatrix} 1 & 2 \\ 2 & 2 \end{pmatrix} \quad \text{and} \quad B = \begin{pmatrix} 3 & 1 \\ 1 & 1 \end{pmatrix}$$

we have $f(x_1, x_2) = \mathbf{x}^T A\mathbf{x}$ and $g(x_1, x_2) = \mathbf{x}^T B\mathbf{x} = 2$. The critical points corresponding to $\lambda_1 = 2$ were $(0, \pm\sqrt{2})$. We have $f(0, \pm\sqrt{2}) = 4$ and $g(0, \pm\sqrt{2}) = 2$. Thus, $f(0, \pm\sqrt{2}) = 2g(0, \pm\sqrt{2}) = 4$. The critical points corresponding to $\lambda_2 = -\frac{1}{2}$ were $(\pm 1, \mp 1)$. We have $f(\pm 1, \mp 1) = -1$ and $g(\pm 1, \mp 1) = 2$. Thus, $f(\pm 1, \mp 1) = -\frac{1}{2}g(\pm 1, \mp 1) = -1$. ∎

13.2 Functions of Three Variables

If $f(x, y)$ is a function of three variables and if $g(x, y, z) = c$ is a constraint then the critical points of $f(x, y, z)$ subject to this constraint are the points (x, y, z) which satisfy the Lagrange multiplier equations $\nabla f = \lambda \nabla g$ and $g(x, y, z) = c$.

For a symmetric 3×3 matrix A its eigenvalue problem is also a special case of a Lagrange multiplier problem with a quadratic function $f(x, y, z) = \mathbf{x}^T A\mathbf{x}$ and a quadratic restraint $g(x, y, z) = \mathbf{x}^T\mathbf{x} = c > 0$ where $\mathbf{x} = \langle x, y, z\rangle^T$. The following specific example will illustrate this.

Example 13.5

Use Lagrange multipliers to find the critical points of

$$\begin{aligned} f(x, y, z) &= 5x^2 - 2xy - 4xz + 8y^2 + 2yz + 5z^2 \\ &= \begin{pmatrix} x & y & z \end{pmatrix} \begin{pmatrix} 5 & -1 & -2 \\ -1 & 8 & 1 \\ -2 & 1 & 5 \end{pmatrix} \begin{pmatrix} x \\ y \\ z \end{pmatrix} \end{aligned}$$

subject to the constraints $g(x, y, z) = x^2 + y^2 + z^2 = 1$ and x is positive, i.e. $0 < x$.

Solution.

The Lagrange multiplier equations are

$$\begin{aligned} \nabla f &= \langle 10x - 2y - 4z, -2x + 16y + 2z, -4x + 2y + 10z\rangle \\ &= \lambda\langle 2x, 2y, 2z\rangle = \lambda \nabla g \end{aligned}$$

and $g(x, y, z) = x^2 + y^2 + z^2 = 1$.

After simplification we have

$$\langle 5x - y - 2z, -x + 8y + z, -2x + y + 5z\rangle = \lambda\langle x, y, z\rangle$$

which is equivalent to the system

$$\begin{aligned} (5-\lambda)x - y - 2z &= 0 \\ -x + (8-\lambda)y + z &= 0 \\ -2x + y + (5-\lambda)z &= 0 \end{aligned}$$

where $x^2 + y^2 + z^2 = 1$. In matrix form the system is given by

$$\begin{pmatrix} 5-\lambda & -1 & -2 \\ -1 & 8-\lambda & 1 \\ -2 & 1 & 5-\lambda \end{pmatrix} \begin{pmatrix} x \\ y \\ z \end{pmatrix} = \begin{pmatrix} 0 \\ 0 \\ 0 \end{pmatrix}$$

where $[x, y, z]^T$ is a unit vector with $x > 0$. But this is just an eigenvalue problem for the matrix

$$A = \begin{pmatrix} 5 & -1 & -2 \\ -1 & 8 & 1 \\ -2 & 1 & 5 \end{pmatrix}$$

where $[x, y, z]^T$ is a unit eigenvector of A with $x > 0$ and the Lagrange multiplier λ is the corresponding eigenvalue. If $\mathbf{x} = \langle x, y, z \rangle^T$ is a unit eigenvector of A with $x > 0$ then $f(x, y, z) = \mathbf{x}^T(A\mathbf{x}) = \mathbf{x}^T(\lambda\mathbf{x}) = \lambda\mathbf{x}^T\mathbf{x} = \lambda$. On the unit sphere $x^2 + y^2 + z^2 = 1$ the function $f(x, y, z)$ will achieve its maximum and minimum values, and $\lambda_{\text{min}} \leq f(x, y, z) \leq \lambda_{\text{max}}$. We need to find the eigenvalues of A and the corresponding unit eigenvectors $\mathbf{x}$ with $x > 0$.

We need to find the characteristic equation for A and will use its principal minors to find c_1, c_2, and c_3 as follows:

$$c_1 = \text{Tr}\, A = 5 + 8 + 5 = 18$$

$$c_2 = \begin{vmatrix} 5 & -1 \\ -1 & 8 \end{vmatrix} + \begin{vmatrix} 5 & -2 \\ -2 & 5 \end{vmatrix} + \begin{vmatrix} 8 & 1 \\ 1 & 5 \end{vmatrix} = 39 + 21 + 39 = 99$$

$$c_3 = \det A = \begin{vmatrix} 5 & -1 & -2 \\ -1 & 8 & 1 \\ -2 & 1 & 5 \end{vmatrix} = \begin{vmatrix} 0 & 39 & 3 \\ -1 & 8 & 1 \\ 0 & -15 & 3 \end{vmatrix} = \begin{vmatrix} 39 & 3 \\ -15 & 3 \end{vmatrix} = 9 \begin{vmatrix} 13 & 1 \\ -5 & 1 \end{vmatrix} = 162 \quad .$$

Thus, $f_A(\lambda) = \lambda^3 - c_1\lambda^2 + c_2\lambda - c_3 = \lambda^3 - 18\lambda^2 + 99\lambda - 162 = (\lambda - 9)(\lambda - 6)(\lambda - 3) = 0$ is the characteristic equation for A and its eigenvalues are $\lambda_1 = 9$, $\lambda_2 = 6$, and $\lambda_3 = 3$.

Corresponding to $\lambda_1 = 9$ we have

$$A - 9I = \begin{pmatrix} -4 & -1 & -2 \\ -1 & -1 & 1 \\ -2 & 1 & -4 \end{pmatrix} \sim \begin{pmatrix} 1 & 1 & -1 \\ 0 & 3 & -6 \\ 0 & 3 & -6 \end{pmatrix} \sim \begin{pmatrix} 1 & 0 & 1 \\ 0 & 1 & -2 \\ 0 & 0 & 0 \end{pmatrix} .$$

Thus,

$$\begin{aligned} x + z &= 0 \\ y - 2z &= 0 . \end{aligned}$$

Set $z = a$ then $[x, y, z]^T = [-a, 2a, a]^T$. Put this into $1 = x^2 + y^2 + z^2 = a^2 + 4a^2 + a^2 = 6a^2$ and use $x > 0$ to obtain $a = -\dfrac{1}{\sqrt{6}}$ and $\langle x, y, z\rangle = \left\langle \dfrac{1}{\sqrt{6}}, -\dfrac{2}{\sqrt{6}}, -\dfrac{1}{\sqrt{6}} \right\rangle$. Then $f\left(\dfrac{1}{\sqrt{6}}, -\dfrac{2}{\sqrt{6}}, -\dfrac{1}{\sqrt{6}}\right) = 9$ is the maximum value of f with the constraint $g(x, y, z) = 1$.

Corresponding to $\lambda_2 = 6$ we have

$$A - 6I = \begin{pmatrix} -1 & -1 & -2 \\ -1 & 2 & 1 \\ -2 & 1 & -1 \end{pmatrix} \sim \begin{pmatrix} 1 & 1 & 2 \\ 0 & 3 & 3 \\ 0 & 0 & 0 \end{pmatrix} \sim \begin{pmatrix} 1 & 0 & 1 \\ 0 & 1 & 1 \\ 0 & 0 & 0 \end{pmatrix} .$$

Thus,

$$\begin{aligned} x + z &= 0 \\ y + z &= 0 . \end{aligned}$$

Set $z = b$ then $[x, y, z]^T = [-b, -b, b]^T$. Put this into $1 = x^2 + y^2 + z^2 = b^2 + b^2 + b^2 = 3b^2$ and use $x > 0$ to obtain $b = -\dfrac{1}{\sqrt{3}}$ and $\langle x, y, z\rangle = \left\langle \dfrac{1}{\sqrt{3}}, \dfrac{1}{\sqrt{3}}, -\dfrac{1}{\sqrt{3}} \right\rangle$. Then $f\left(\dfrac{1}{\sqrt{3}}, \dfrac{1}{\sqrt{3}}, -\dfrac{1}{\sqrt{3}}\right) = 6$.

Corresponding to $\lambda_3 = 3$ we have

$$A - 3I = \begin{pmatrix} 2 & -1 & -2 \\ -1 & 5 & 1 \\ -2 & 1 & 2 \end{pmatrix} \sim \begin{pmatrix} 1 & -5 & -1 \\ 0 & 9 & 0 \\ 0 & 0 & 0 \end{pmatrix} \sim \begin{pmatrix} 1 & 0 & -1 \\ 0 & 1 & 0 \\ 0 & 0 & 0 \end{pmatrix} .$$

Thus,

$$\begin{aligned} x - z &= 0 \\ y &= 0 \quad . \end{aligned}$$

Set $z = c$ then $[x, y, z]^T = [c, 0, c]$. Put this into $1 = x^2 + y^2 + z^2 = c^2 + c^2 = 2c^2$ and use $x > 0$ to obtain $c = \dfrac{1}{\sqrt{2}}$ and $\langle x, y, z\rangle = \left\langle \dfrac{1}{\sqrt{2}}, 0, \dfrac{1}{\sqrt{2}} \right\rangle$. Then $f\left(\dfrac{1}{\sqrt{2}}, 0, \dfrac{1}{\sqrt{2}}\right) = 3$ is the minimum value of f with the constraint $g(x, y, z) = 1$. ∎

13.3 The General Case: Functions of n Variables

If $f = f(x_1, \ldots, x_n)$ is a function of n variables and if $g(x_1, \ldots, x_n) = c$ is a constraint then the critical points of $f(x_1, \ldots, x_n)$ subject to this constraint are the points $\mathbf{x} = (x_1, \ldots, x_n)$ which satisfy the Lagrange multiplier equations

$$\nabla f = \lambda \nabla g \quad \text{and} \quad g(x_1, \ldots, x_n) = c \quad .$$

In general if A is an $n \times n$ symmetric matrix and if $f(\mathbf{x}) = \mathbf{x}^T A\mathbf{x}$ where $g(\mathbf{x}) = \mathbf{x}^T\mathbf{x} = c > 0$ then the critical points of f are eigenvectors $\mathbf{x} = [x_1, \ldots, x_n]^T$ of length $\sqrt{c}$, the Lagrange multiplier λ is the corresponding eigenvalue, and $f(\mathbf{x}) = \lambda c$. This was shown in Example 13.1 for $n = 2$. Furthermore, on $x_1^2 + \cdots + x_n^2 = c > 0$ the function $f(\mathbf{x})$ achieve its maximum and minimum values, and $\lambda_{\min} \le f(\mathbf{x}) \le \lambda_{\max}$.

Let B be an $n \times n$ symmetric matrix and let the constraint function $g(\mathbf{x}) = \mathbf{x}^T B\mathbf{x}$. Then in general the critical points of $f(\mathbf{x}) = \mathbf{x}^T A\mathbf{x}$ subject to the constraint $g(\mathbf{x}) = \mathbf{x}^T B\mathbf{x} = c \neq 0$ are the solutions $\mathbf{x}$ to the system $(A - \lambda B)\mathbf{x} = 0$ which satisfy $g(\mathbf{x}) = \mathbf{x}^T B\mathbf{x} = c$. Each solution $\mathbf{x}$ corresponds to a λ for which the polynomial $\det(\lambda B - A) = 0$. Example 13.3 shows this for $n = 2$.

13.4 Exercises

1. Use Lagrange multipliers to find the critical points of $f(x_1, x_2) = \mathbf{x}^T A\mathbf{x} = a_{11}x_1^2 + 2a_{12}x_1x_2 + a_{22}x_2^2$ with $x_1 > 0$ subject to the constraint $g(x_1, x_2) = \mathbf{x}^T\mathbf{x} = x_1^2 + x_2^2 = c$ if

   a) $A = \begin{pmatrix} 6 & 2 \\ 2 & 9 \end{pmatrix}$ and $c = 20$

   b) $A = \begin{pmatrix} -10 & 6 \\ 6 & -5 \end{pmatrix}$ and $c = 13$

c) $A = \begin{pmatrix} 7 & 4 \\ 4 & 13 \end{pmatrix}$ and $c = 5$

d) $A = \begin{pmatrix} 9 & -4 \\ -4 & 9 \end{pmatrix}$ and $c = 2$.

2. Use Lagrange multipliers to find the critical points of

$$f(x_1, x_2) = \mathbf{x}^T A \mathbf{x} = a_{11}x_1^2 + a_{12}x_1x_2 + a_{22}x_2^2$$

with $x_1 > 0$ subject to the constraint $g(x_1, x_2) = \mathbf{x}^T B \mathbf{x} = b_{11}x_1^2 + 2b_{12}x_1x_2 + b_{22}x_2^2 = c$ if

a) $A = \begin{pmatrix} 0 & 4 \\ 4 & 0 \end{pmatrix}$, $B = \begin{pmatrix} 1 & 0 \\ 0 & 4 \end{pmatrix}$, and $c = 8$

b) $A = \begin{pmatrix} 1 & 2 \\ 2 & 1 \end{pmatrix}$, $B = \begin{pmatrix} 2 & 1 \\ 1 & 2 \end{pmatrix}$, and $c = 6$

c) $A = \begin{pmatrix} 2 & 3 \\ 3 & -12 \end{pmatrix}$, $B = \begin{pmatrix} 1 & 0 \\ 0 & -1 \end{pmatrix}$, and $c = 10$

d) $A = \begin{pmatrix} 12 & 4 \\ 4 & 1 \end{pmatrix}$, $B = \begin{pmatrix} 4 & 8 \\ 8 & 8 \end{pmatrix}$, and $c = 56$

e) $A = \begin{pmatrix} 1 & 1 \\ 1 & 1 \end{pmatrix}$, $B = \begin{pmatrix} 0 & 1 \\ 1 & 0 \end{pmatrix}$, and $c = -8$

3. Use Lagrange multipliers to find the critical points of $f(\mathbf{x}) = \mathbf{x}^T A \mathbf{x}$ with $x_1 > 0$ subject to the constraint $g(\mathbf{x}) = \mathbf{x}^T \mathbf{x} = c$ if

a) $A = \begin{pmatrix} 3 & 2 & -1 \\ 2 & 3 & -1 \\ -1 & -1 & 6 \end{pmatrix}$ and $c = 6$

b) $A = \begin{pmatrix} 7 & -2 & 1 \\ -2 & 10 & -2 \\ 1 & -2 & 7 \end{pmatrix}$ and $c = 30$ (See Chapter 7, Exercise 19, p. 115)

c) $A = \begin{pmatrix} 3 & -1 & 0 \\ -1 & 2 & -1 \\ 0 & -1 & 3 \end{pmatrix}$ and $c = 12$

d) $A = \begin{pmatrix} 7 & -2 & 4 \\ -2 & 4 & -2 \\ 4 & -2 & 7 \end{pmatrix}$ and $c = 18$ (See Chapter 7, Exercise 20, p. 115)

4. Use Lagrange multipliers to find the critical points of $f(\mathbf{x}) = \mathbf{x}^T A\mathbf{x}$ with $x_1 > 0$ subject to the constraints $g(\mathbf{x}) = \mathbf{x}^T B\mathbf{x} = c$ if $f(\mathbf{x}) \neq 0$ and if

a) $A = \begin{pmatrix} 1 & 1 & 1 \\ 1 & 1 & 1 \\ 1 & 1 & 1 \end{pmatrix}$, $B = \begin{pmatrix} 2 & 0 & 0 \\ 0 & 2 & 0 \\ 0 & 0 & 1 \end{pmatrix}$, and $c = 8$

b) $A = \begin{pmatrix} 2 & 0 & 0 \\ 0 & 0 & 1 \\ 0 & 1 & 0 \end{pmatrix}$, $B = \begin{pmatrix} 1 & 0 & 0 \\ 0 & -1 & 0 \\ 0 & 0 & 1 \end{pmatrix}$, and $c = 4$.

A Answers to the Exercises

A.1 Chapter 1

1. $\begin{pmatrix} 2 & -2 & 7 \\ 3 & -4 & 9 \end{pmatrix}$.

2. $\begin{pmatrix} 3 & 2 \\ 7 & 5 \\ 7 & 8 \end{pmatrix}$.

3. $\begin{pmatrix} 6 & 3 & 3 & 15 \\ 3 & 0 & 3 & 9 \\ 6 & 12 & 3 & 15 \end{pmatrix}$.

4. $\begin{pmatrix} \alpha & 6\alpha & 5\alpha \\ 3\alpha & 0 & \alpha \\ 2\alpha & \alpha & \alpha \end{pmatrix}$.

5. $\begin{pmatrix} \alpha & 5\alpha & 3\alpha & 2\alpha \\ 0 & \beta & \beta & \beta \end{pmatrix}$.

6. $\begin{pmatrix} 6 & 0 \\ -4 & 0 \end{pmatrix}$.

7. $\begin{pmatrix} 3 & 2 & 5 \\ -2 & 4 & 6 \end{pmatrix}$.

8. $\begin{pmatrix} 15 & 14 \end{pmatrix}$.

9. $\begin{array}{rclcrcl} a & = & 1 & , & b & = & 3 \\ c & = & -4 & , & d & = & -11 \end{array}$.

10. $\begin{pmatrix} u \\ v \end{pmatrix} = \begin{pmatrix} 8 & 4 \\ 2 & -1 \end{pmatrix} \begin{pmatrix} w \\ x \end{pmatrix}$.

11. $a_{32} = 3 \quad a_{23} = -3$ dimensions are is 3×4 $\quad \mathbf{a}_2 = \begin{pmatrix} -1 \\ 2 \\ 3 \end{pmatrix}$

$\mathbf{a}_{(3)} = \begin{pmatrix} 5 & 3 & -1 & 2 \end{pmatrix} \quad A^T = \begin{pmatrix} 1 & 0 & 5 \\ -1 & 2 & 3 \\ 2 & -3 & -1 \\ 3 & 1 & 2 \end{pmatrix}$.

12. $\mathbf{a} = \begin{pmatrix} 1 \\ -1 \\ 3 \end{pmatrix}$ $\mathbf{b} = \begin{pmatrix} 2 \\ 1 \\ -1 \end{pmatrix}$ $\mathbf{a} \bullet \mathbf{b} = \mathbf{a}^T\mathbf{b} = \begin{pmatrix} 1 & -1 & 3 \end{pmatrix} \begin{pmatrix} 2 \\ 1 \\ -1 \end{pmatrix} =$ -2 .

13. $X = \begin{pmatrix} 3 & -1 & 2 \\ 2 & 1 & 4 \end{pmatrix}$.

14. **a)** $\begin{pmatrix} 5 & 6 & 7 \\ -2 & 0 & 2 \\ 2 & 0 & -2 \end{pmatrix}$

b) $\begin{pmatrix} 3 \end{pmatrix}$.

15. **a)** $\left(\begin{array}{cc|cc} 3 & -1 & 4 & 5 \\ 2 & 4 & 5 & -6 \end{array}\right)$ **b)** $\left(\begin{array}{ccc|ccc} 1 & 1 & 1 & 1 & 4 & 5 \\ 2 & 2 & 2 & -4 & -4 & 4 \\ 4 & 4 & 4 & -3 & -1 & 7 \end{array}\right)$.

A.2 Chapter 2

1. **a)** $\begin{pmatrix} 2 & 3 & 5 & 7 \end{pmatrix}$, $\left(\begin{array}{cccc|c} 2 & 3 & 5 & 7 & 8 \end{array}\right)$ **b)** $\begin{pmatrix} 3 & 2 \\ 4 & 3 \\ 5 & 4 \\ 6 & 5 \end{pmatrix}$, $\left(\begin{array}{cc|c} 3 & 2 & 1 \\ 4 & 3 & 2 \\ 5 & 4 & 3 \\ 6 & 5 & 4 \end{array}\right)$

c) $\begin{pmatrix} 2 & 2 & -1 \\ 1 & 1 & 2 \\ 1 & 1 & 1 \end{pmatrix}$, $\left(\begin{array}{ccc|c} 2 & 2 & -1 & -1 \\ 1 & 1 & 2 & 2 \\ 1 & 1 & 1 & 1 \end{array}\right)$

d) $\begin{pmatrix} 2 & 1 & -4 & 4 \\ 1 & 2 & 1 & 4 \end{pmatrix}$, $\left(\begin{array}{cccc|c} 2 & 1 & -4 & 4 & 1 \\ 1 & 2 & 1 & 4 & 2 \end{array}\right)$

e) $\begin{pmatrix} 1 & 2 \\ 1 & 3 \\ 2 & 5 \end{pmatrix}$, $\left(\begin{array}{cc|c} 1 & 2 & 0 \\ 1 & 3 & 0 \\ 2 & 5 & 0 \end{array}\right)$ **f)** $\begin{pmatrix} 3 & 5 \\ 2 & 7 \end{pmatrix}$, $\left(\begin{array}{cc|c} 3 & 5 & 4 \\ 2 & 7 & 1 \end{array}\right)$.

2. **a)** $\begin{pmatrix} 1 & 0 & -1 & -5 \\ 0 & 1 & 1 & 3 \end{pmatrix}$ **b)** $\begin{pmatrix} 1 & 0 & 0 & 0 \\ 0 & 1 & 0 & 0 \\ 0 & 0 & 1 & 1 \end{pmatrix}$ **c)** $\begin{pmatrix} 1 & 0 & 0 \\ 0 & 1 & 0 \\ 0 & 0 & 1 \end{pmatrix}$

d) $\begin{pmatrix} 1 & 1 & 0 & 0 \\ 0 & 0 & 1 & 1 \\ 0 & 0 & 0 & 0 \end{pmatrix}$ **e)** $\begin{pmatrix} 1 & 1 & 0 & 0 \\ 0 & 0 & 1 & 0 \\ 0 & 0 & 0 & 1 \end{pmatrix}$ **f)** $\begin{pmatrix} 0 & 1 & 0 & 2 \\ 0 & 0 & 1 & 3 \end{pmatrix}$.

3. **a)** $\begin{array}{rcl} x_1 & = & 0 \\ x_2 & = & 1 \end{array}$ **b)** $\begin{array}{rcl} x_1 & = & 1 \\ 2x_1 & = & 0 \end{array}$ **c)** $\begin{array}{rcl} x_1 + 2x_2 - x_3 & = & 1 \\ 3x_1 - \quad x_2 \quad\quad & = & 2 \\ x_2 + x_3 & = & -1 \end{array}$

d) $\begin{array}{rcl} x_1 - x_2 & = & 2 \\ 2x_1 \quad\quad & = & -1 \\ x_1 + x_2 & = & 0 \end{array}$.

4. First matrix: Condition 2. Second matrix: Condition 1. Third matrix: Condition 3, Fourth matrix: Condition 4.

5. Find the full row reduced echelon form for each system. Since there is only one such form for a system, the two systems are equivalent if and only if they reduce to the same echelon form. In this case, they do, so the systems are equivalent. Their row reduced echelon form is

$$\left(\begin{array}{cccc|c} 1 & 0 & 0 & 0 & -1 \\ 0 & 1 & 0 & 0 & 1 \\ 0 & 0 & 1 & 0 & 0 \\ 0 & 0 & 0 & 1 & -1 \end{array}\right) .$$

6.

$$\left(\begin{array}{cccc} 0 & 1 & 0 & 3 \\ 0 & 0 & 1 & 2 \\ 0 & 0 & 0 & 0 \end{array}\right) .$$

7.

$$R = \left(\begin{array}{ccccc} 0 & 0 & 1 & 0 & 0 \\ 0 & 0 & 0 & 1 & 0 \\ 0 & 0 & 0 & 0 & 1 \end{array}\right) .$$

A.3 Chapter 3

1. When the augmented matrix of an inconsistent system is row reduced into row reduced echelon form, there is a row with an isolated non-zero element at the extreme right with zeros preceding it.

2. **a)** $\left(2 \right)$ **b)** $\left(\begin{array}{c} 1 \\ -1 \end{array}\right)$ **c)** $\left(\begin{array}{c} 0 \\ 0 \end{array}\right)$ **d)** $\left(\begin{array}{c} 5 \\ -2 \end{array}\right)$ **e)** $\left(\begin{array}{c} 0 \\ 0 \\ 0 \end{array}\right)$ **f)** $\left(\begin{array}{c} 1 \\ 1 \\ -1 \end{array}\right)$

g) $\begin{pmatrix} -2 \\ 0 \\ 1 \end{pmatrix}$ **h)** $\begin{pmatrix} 1 \\ 0 \\ 1 \\ -1 \end{pmatrix}$ **i)** $\begin{pmatrix} 0 \\ 0 \\ 0 \\ 0 \end{pmatrix}$ **j)** $\begin{pmatrix} 0 \\ -1 \\ 2 \\ 0 \end{pmatrix}$.

3. **a)** $\begin{pmatrix} 1 \\ 0 \end{pmatrix} + a \begin{pmatrix} -1 \\ 1 \end{pmatrix}$ **b)** $a \begin{pmatrix} 2 \\ 1 \end{pmatrix}$ **c)** $\begin{pmatrix} 2 \\ 0 \\ 0 \end{pmatrix} + a \begin{pmatrix} -1 \\ 1 \\ 0 \end{pmatrix} + b \begin{pmatrix} 3 \\ 0 \\ 1 \end{pmatrix}$

d) $\begin{pmatrix} \frac{17}{5} \\ \frac{-1}{5} \\ 0 \end{pmatrix} + a \begin{pmatrix} \frac{-2}{5} \\ \frac{1}{5} \\ 1 \end{pmatrix}$ **e)** $\begin{pmatrix} \frac{1}{2} \\ \frac{-1}{2} \\ 0 \end{pmatrix} + a \begin{pmatrix} \frac{-1}{2} \\ \frac{-1}{2} \\ 1 \end{pmatrix}$

f) $\begin{pmatrix} 1 \\ -2 \\ 0 \\ 0 \end{pmatrix} + a \begin{pmatrix} 0 \\ 1 \\ 1 \\ 0 \end{pmatrix} + b \begin{pmatrix} -1 \\ 0 \\ 0 \\ 1 \end{pmatrix}$ **g)** $a \begin{pmatrix} \frac{-1}{2} \\ \frac{1}{2} \\ -1 \\ 1 \\ 0 \end{pmatrix} + b \begin{pmatrix} \frac{1}{2} \\ \frac{1}{2} \\ -1 \\ 0 \\ 1 \end{pmatrix}$.

4. **a)** None **b)** $\begin{pmatrix} 2 \\ -1 \\ 0 \end{pmatrix}$ **c)** $\begin{pmatrix} 2 \\ 0 \end{pmatrix} + a \begin{pmatrix} -3 \\ 1 \end{pmatrix}$ **d)** $\begin{pmatrix} 1 \\ 0 \\ -1 \end{pmatrix}$

e) $a \begin{pmatrix} 1 \\ 1 \\ -1 \\ 0 \\ 1 \end{pmatrix} + b \begin{pmatrix} 1 \\ -1 \\ 0 \\ 1 \\ 0 \end{pmatrix}$ **f)** $\begin{pmatrix} 2 \\ -1 \\ 1 \end{pmatrix}$.

5. $2x + 3y - z = 1$.

6. **a)** No solution **b)** $\begin{pmatrix} -2\alpha \\ \alpha \end{pmatrix} + \begin{pmatrix} -1 \\ 0 \end{pmatrix}$ **c)** No solution

d) $\begin{pmatrix} 2\alpha \\ \alpha \\ \alpha \end{pmatrix} + \begin{pmatrix} 3 \\ 0 \\ 0 \end{pmatrix}$ **e)** $\begin{pmatrix} 2\alpha \\ -3\alpha \\ \alpha \\ \alpha \end{pmatrix} + \begin{pmatrix} -3 \\ 8 \\ -3 \\ 0 \end{pmatrix}$ **f)** No solution .

7.

$$\begin{pmatrix} x_1 \\ x_2 \\ x_3 \\ x_4 \\ x_5 \\ x_6 \\ x_7 \end{pmatrix} = \begin{pmatrix} 0 \\ 2 \\ 0 \\ 8 \\ 0 \\ 9 \\ 0 \end{pmatrix} + \alpha \begin{pmatrix} 1 \\ 0 \\ 0 \\ 0 \\ 0 \\ 0 \\ 0 \end{pmatrix} + \beta \begin{pmatrix} 0 \\ 0 \\ 1 \\ 0 \\ 0 \\ 0 \\ 0 \end{pmatrix} + \gamma \begin{pmatrix} 0 \\ -3 \\ 0 \\ -5 \\ 1 \\ 0 \\ 0 \end{pmatrix} + \delta \begin{pmatrix} 0 \\ -7 \\ 0 \\ -6 \\ 0 \\ -4 \\ 1 \end{pmatrix} .$$

A.4 Chapter 4

1. **a)** 14 **b)** -1 **c)** $\ln(4/27)$ **d)** 1 **e)** $\tan^2 x$ **f)** 17 .

2. **a)** 90 **b)** 0 **c)** -5 **d)** -24 **e)** -1 **f)** 10 .

3. **a)** $-1/2$ **b)** $1, -3$ **c)** $0, 2, 3$ **d)** $1, 2, -2$.

4. **a)** Hint: expand the determinant along the bottom row.

b) Hint: Use the fact stated in (4.19).

c) $2x + 3y = 1$.

5. $A = 1/2$.

6. $V = 1$.

7. Hint: Use Theorem 4.3.

8. **a)** $\begin{matrix} x = 1 \\ y = 2 \end{matrix}$ **b)** $\begin{matrix} x = \frac{1}{2} \\ y = \frac{1}{3} \end{matrix}$ **c)** $\begin{matrix} x = -3 \\ y = -1 \end{matrix}$ **d)** $\begin{matrix} x = 0 \\ y = -1 \end{matrix}$.

9. 4.

10. 35 and -35.

11. **a)** -13 **b)** -4 **c)** 0 **d)** -6 **e)** 12 **f)** -2 .

A.5 Chapter 5

1. Inverses exist only for matrices in parts **b)**, **d)**, **f)**, **g)**, **j)**, **l)**, and **m)** .

2. **b)** $\left(\begin{array}{c} \frac{1}{2} \end{array}\right)$ **d)** $\frac{1}{3}\left(\begin{array}{rr} -1 & 2 \\ 2 & -1 \end{array}\right)$ **i)** $\left(\begin{array}{cc} 0 & 1 \\ 1 & 0 \end{array}\right)$

g) $\left(\begin{array}{cc} 1 & 0 \\ 0 & 1 \end{array}\right)$ **j)** $\frac{1}{4}\left(\begin{array}{rrr} 3 & -1 & 1 \\ 1 & 1 & 3 \\ 1 & 1 & -1 \end{array}\right)$ **l)** $\left(\begin{array}{cccc} 0 & 0 & 0 & 1 \\ 0 & 0 & 1 & 0 \\ 0 & 1 & 0 & 0 \\ 1 & 0 & 0 & 0 \end{array}\right)$ **m)** $\left(\begin{array}{cccc} 1 & 0 & 0 & 0 \\ 0 & 1 & 0 & 0 \\ 0 & 0 & 1 & 0 \\ 0 & 0 & 0 & 1 \end{array}\right)$

3. Yes.

4. $ad - bc \neq 0,\ A^{-1} = \dfrac{1}{ad - bc}\left(\begin{array}{rr} d & -b \\ -c & a \end{array}\right)$.

5. $abc \neq 0,\ A^{-1} = \left(\begin{array}{ccc} \frac{1}{a} & 0 & 0 \\ 0 & \frac{1}{b} & 0 \\ 0 & 0 & \frac{1}{c} \end{array}\right)$.

6. **a)** $\left(\begin{array}{r} 0 \\ -1 \\ 1 \end{array}\right)$ **b)** $\left(\begin{array}{r} 5 \\ -2 \\ -6 \end{array}\right)$ **c)** $\left(\begin{array}{c} 0 \\ 0 \\ 1 \end{array}\right)$ **d)** $\left(\begin{array}{c} 0 \\ 0 \\ 0 \end{array}\right)$ **e)** $\left(\begin{array}{r} 13 \\ -10 \\ -11 \end{array}\right)$.

7. $x = 13,\ y = -10$, and $z = -11$.

8. $A^{-1} = \left(\begin{array}{rrr} 0 & -1 & 1 \\ 0 & 0 & 1 \\ 1 & 1 & -1 \end{array}\right)$.

9. $A^{-1} = \left(\begin{array}{cc} \cos\beta & \sin\beta \\ -\sin\beta & \cos\beta \end{array}\right)$, $A^{-1} = A^T$ & $A^{-1}(\beta) = A(-\beta)$.

10. **a)** Division by zero would occur. **b)** $\left(\begin{array}{c} x \\ y \end{array}\right) = \left(\begin{array}{c} 3 \\ 0 \end{array}\right) + \alpha\left(\begin{array}{r} -1 \\ 1 \end{array}\right)$.

11. **a)** $P = \left(\begin{array}{ccc} 1 & 0 & 0 \\ 0 & 1 & 0 \\ \beta & 0 & 1 \end{array}\right)$

b) $P = \left(\begin{array}{cccc} 1 & 0 & 0 & 0 \\ 0 & 0 & 0 & 1 \\ 0 & 0 & 1 & 0 \\ 0 & 1 & 0 & 0 \end{array}\right)$

c) $P = \left(\begin{array}{cccc} 1 & 0 & 0 & 0 \\ 0 & \alpha & 0 & 0 \\ 0 & 0 & 1 & 0 \\ 0 & 0 & 0 & 1 \end{array}\right)$.

12. **a)** $P = \begin{pmatrix} 1 & 0 & 0 \\ 0 & 1 & 0 \\ -\beta & 0 & 1 \end{pmatrix}$

b) $P = \begin{pmatrix} 1 & 0 & 0 & 0 \\ 0 & 0 & 0 & 1 \\ 0 & 0 & 1 & 0 \\ 0 & 1 & 0 & 0 \end{pmatrix}$

c) $P = \begin{pmatrix} 1 & 0 & 0 & 0 \\ 0 & \frac{1}{\alpha} & 0 & 0 \\ 0 & 0 & 1 & 0 \\ 0 & 0 & 0 & 1 \end{pmatrix}$.

13. **a)** $A = \begin{pmatrix} 7 & 2 \\ 3 & 1 \end{pmatrix}, \quad A^{-1} = \begin{pmatrix} 1 & -2 \\ -3 & 7 \end{pmatrix}$

b) $A = \begin{pmatrix} 1 & 3 & 0 \\ 0 & 1 & 0 \\ 0 & 2 & 1 \end{pmatrix}, \quad A^{-1} = \begin{pmatrix} 1 & -3 & 0 \\ 0 & 1 & 0 \\ 0 & -2 & 1 \end{pmatrix}$

c) $A = \begin{pmatrix} 0 & 1 & 4 \\ 1 & 0 & 0 \\ 0 & 0 & 1 \end{pmatrix}, \quad A^{-1} = \begin{pmatrix} 0 & 1 & 0 \\ 1 & 0 & -4 \\ 0 & 0 & 1 \end{pmatrix}$

d) $A = \begin{pmatrix} 1 & 0 & 0 & 0 \\ 15 & 1 & 3 & 0 \\ 5 & 0 & 1 & 0 \\ 0 & 0 & 0 & 1 \end{pmatrix}, \quad \begin{pmatrix} 1 & 0 & 0 & 0 \\ 0 & 1 & -3 & 0 \\ -5 & 0 & 1 & 0 \\ 0 & 0 & 0 & 1 \end{pmatrix}$.

14. **a)** $\begin{pmatrix} 7 & 0 \\ 0 & 2 \end{pmatrix}$ **b)** $\begin{pmatrix} 0 & -2 \\ -3 & 0 \end{pmatrix}$ **c)** $\begin{pmatrix} 7 & -3 \\ 5 & 0 \end{pmatrix}$ **d)** $\begin{pmatrix} 7 & -3 \\ 5 & 2 \end{pmatrix}$

e) $\begin{pmatrix} 1 & 5 & 2 \\ 2 & -7 & 4 \\ 7 & 1 & -3 \end{pmatrix}$ **f)** $\begin{pmatrix} -2 & 0 & 2 \\ 0 & 0 & -1 \\ -2 & -2 & 2 \end{pmatrix}$ **g)** $\begin{pmatrix} 2 & 0 & 0 \\ 0 & 8 & -10 \\ 0 & -4 & 6 \end{pmatrix}$.

16. **a)** $\begin{pmatrix} -1 \\ 2 \end{pmatrix}$ **b)** $\begin{pmatrix} 1 & 2 \\ 1 & -2 \\ -4 & 5 \end{pmatrix}$ **c)** $\begin{pmatrix} 3 & 7 & -3 & 5 \\ -3 & -1 & 6 & 1 \\ 6 & -7 & 9 & 1 \end{pmatrix}$

d) $\begin{pmatrix} 1 \\ -7 \\ 1 \\ 1 \end{pmatrix}$.

18. **a)** $\begin{pmatrix} 32 & 39 \\ -41 & 50 \end{pmatrix}$ **b)** $\frac{1}{4}\begin{pmatrix} 2 & 4 & 2 \\ 3 & 2 & 1 \\ 2 & 0 & 2 \end{pmatrix}$ **c)** $\begin{pmatrix} 1 & -3 & 1 \\ 0 & 1 & 0 \\ -2 & 7 & 3 \end{pmatrix}$.

A.6 Chapter 6

1. Only the matrices in **b)**, **c)**, **d)**, and **e)** are orthogonal.

2. **a)** $(1,0,0)' : (1/3,-2/3,2/3)$, $(0,1,0)' : (2/3,2/3,1/3)$, $(0,0,1)' : (-2/3,1/3,2/3)$
b) $(6,0,3) : (4,5,-2)'$ **c)** $(6,0,3)' : (0,-3,6)$.

3. **a)** $(1,0,0)' : (1/\sqrt{2},1/\sqrt{2},0)$, $(0,1,0)' : (-1/\sqrt{2},1/\sqrt{2},0)$, $(0,0,1)' : (0,0,1)$
b) $(1,1,1) : (\sqrt{2},0,1)'$ **c)** $(1,1,1)' : (0,\sqrt{2},1)$.

4. No.

5. $b_1 = c_1 = c_2 = 0$ and the other requirement is $|a_1| = |b_2| = |c_3| = 1$.

$$\therefore \quad Q = \begin{pmatrix} \pm 1 & 0 & 0 \\ 0 & \pm 1 & 0 \\ 0 & 0 & \pm 1 \end{pmatrix} .$$

6. $0 = \mathbf{q}_1^T \mathbf{q}_2 = \frac{1}{\sqrt{6}}(a_1 + 2a_2 - 3a_1) \Longrightarrow 2a_2 - 2a_1 = 0 \Longrightarrow a_2 = a_1$ Then $\mathbf{q}_2 = a_1 \begin{pmatrix} 1 \\ 1 \\ 3 \end{pmatrix}$.

Since $|\mathbf{q}_2| = 1$, $1 = |a_1|\sqrt{1+1+9} = \sqrt{11}\,|a_1|$. Therefore $\mathbf{q}_2 = \pm\frac{1}{\sqrt{11}}\begin{pmatrix} 1 \\ 1 \\ 3 \end{pmatrix}$.

Now $\mathbf{q}_3 = \mathbf{q}_1 \times \mathbf{q}_2$ which is

$$\mathbf{q}_3 = \pm\begin{vmatrix} \mathbf{i} & \mathbf{j} & \mathbf{k} \\ \frac{1}{\sqrt{6}} & \frac{2}{\sqrt{6}} & \frac{-1}{\sqrt{6}} \\ \frac{1}{\sqrt{11}} & \frac{1}{\sqrt{11}} & \frac{3}{\sqrt{11}} \end{vmatrix} = \pm\frac{1}{\sqrt{66}}\langle 7,-4,-1\rangle^T .$$

Hence,

$$\mathbf{q}_3 = \frac{\pm 1}{\sqrt{66}}\begin{pmatrix} 7 \\ -4 \\ -1 \end{pmatrix} .$$

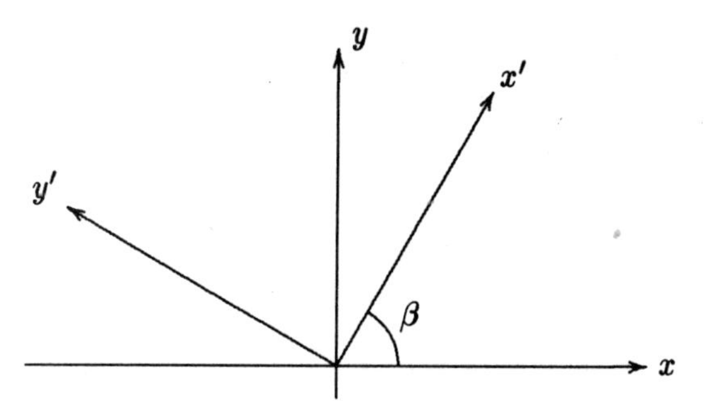

Figure A.1

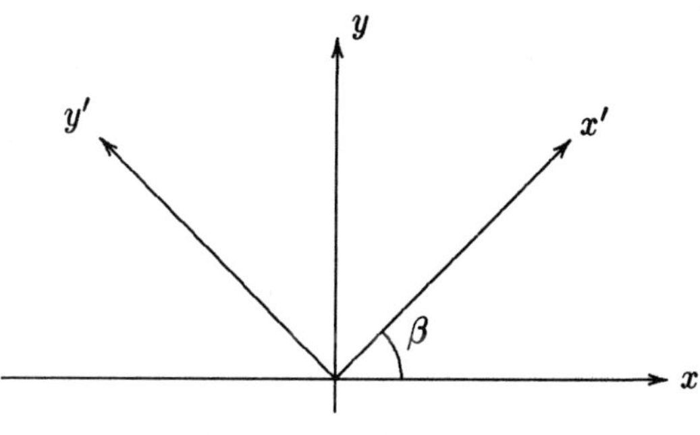

Figure A.2

$$\therefore \qquad Q = \begin{pmatrix} \frac{1}{\sqrt{6}} & \frac{\pm 1}{\sqrt{11}} & \frac{\pm 7}{\sqrt{66}} \\ \frac{2}{\sqrt{6}} & \frac{\pm 1}{\sqrt{11}} & \frac{\mp 4}{\sqrt{66}} \\ \frac{-1}{\sqrt{6}} & \frac{\pm 3}{\sqrt{11}} & \frac{\mp 1}{\sqrt{66}} \end{pmatrix} .$$

Here you must consistently select the upper signs or select the lower signs throughout.

7. **a)** $\beta = \pi/3$.

b) See Figure A.1. The basis vectors are $\mathbf{u}_1 = \langle \frac{1}{2}, \frac{\sqrt{3}}{2} \rangle$ and $\mathbf{u}_2 = \langle \frac{-\sqrt{3}}{2}, \frac{1}{2} \rangle$.

8. **a)** $\beta = \pi/4$.

b) See Figure A.2. The new basis vectors are $\mathbf{u}_1 = \langle \frac{1}{\sqrt{2}}, \frac{1}{\sqrt{2}} \rangle$ and $\mathbf{u}_2 = \langle \frac{-1}{\sqrt{2}}, \frac{1}{\sqrt{2}} \rangle$

9. $Q = \begin{pmatrix} 0 & 0 & 1 \\ 1 & 0 & 0 \\ 0 & -1 & 0 \end{pmatrix}$ No, this is a reflection of coordinate axes since $\det Q = -1$.

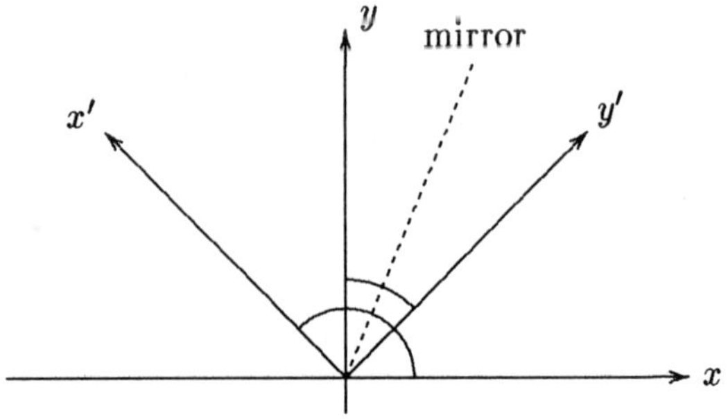

Figure A.3

10. No, $\det Q = -1$ so it corresponds to a reflection of coordinate axes. See figure A.3.

13. $-x' + y' - z' = 6$.

16. **a)** $(4, 4, -3)$ **b)** $(5, 3, -5)$ **c)** $(7, 9, -4)$.

17. **a)** $(0, 1, 5)$ **b)** $(-8, -1, -1)$ **c)** $(-12, -1, 1)$.

18. **a)** $(-5, 7, -4)$ **b)** $(-4, 3, -1)$ **c)** $(0, -7, 8)$.

A.7 Chapter 7

3. **a)** $\lambda^2 - 2\lambda - 5$ **b)** $\lambda^2 - 5\lambda + 7$ **c)** $\lambda^2 + 5\lambda + 1$ **d)** $\lambda^2 + 3\lambda - 13$

e) $\lambda^2 - 9$ **f)** $\lambda^2 + 9$ **g)** $\lambda^3 - 5\lambda^2 + 5\lambda - 16$

h) $\lambda^3 - 4\lambda + 4$ **i)** $\lambda^3 + 2\lambda^2 - 4\lambda - 3$ **j)** $\lambda^3 - 6\lambda^2 + 14\lambda - 2$.

4. and **5.** NOTE: Your eigenvectors may be a non-zero scalar multiples of the answers given below.

a) $0 : \begin{pmatrix} 1 \\ -2 \end{pmatrix}$ $5 : \begin{pmatrix} 2 \\ 1 \end{pmatrix}$

b) $1 : \begin{pmatrix} 1 \\ 2 \end{pmatrix}$ $6 : \begin{pmatrix} -2 \\ 1 \end{pmatrix}$

c) $1 : \begin{pmatrix} -1 \\ 1 \\ 1 \end{pmatrix}$ $4 : \begin{pmatrix} 2 \\ 1 \\ 1 \end{pmatrix}$ $-2 : \begin{pmatrix} 0 \\ -1 \\ 1 \end{pmatrix}$

d) $2: \begin{pmatrix} 1 \\ -1 \\ 0 \end{pmatrix}$ $\quad 3: \begin{pmatrix} 1 \\ 0 \\ 0 \end{pmatrix}$ $\quad 5: \begin{pmatrix} 3 \\ 2 \\ 1 \end{pmatrix}$

e) $1: \begin{pmatrix} 0 \\ 1 \end{pmatrix}$

f) $2: \begin{pmatrix} 1 \\ 1 \end{pmatrix}$

g) $0: \begin{pmatrix} 0 \\ 1 \\ -2 \end{pmatrix}$

$5:$ any non-zero linear combination of $\begin{pmatrix} 1 \\ 0 \\ 0 \end{pmatrix}$ and $\begin{pmatrix} 0 \\ 2 \\ 1 \end{pmatrix}$

h) $3: \begin{pmatrix} 0 \\ 1 \\ -2 \end{pmatrix}$.

$2:$ any non-zero linear combination of $\begin{pmatrix} 1 \\ 0 \\ 0 \end{pmatrix}$ and $\begin{pmatrix} 0 \\ 1 \\ -1 \end{pmatrix}$.

6. The vector $\mathbf{u}$ is not an eigenvector, but $\mathbf{v}$ is an eigenvector corresponding to the eigenvalue 6.

7. **a)** The vector $\mathbf{u}$ is not an eigenvector, but $\mathbf{v}$ is an eigenvector corresponding to the eigenvalue 8.

b) The vector $\mathbf{u}$ is not an eigenvector, but $\mathbf{v}$ is an eigenvector corresponding to the eigenvalue 7.

8. **a)** $\lambda_1 = -1$, $\lambda_2 = 3$, and $\lambda_3 = 8$

b) $\lambda_1 = 11$, $\lambda_2 = 8$, and $\lambda_3 = 3$.

9. **a)** $\mathbf{q}_1 = \frac{1}{\sqrt{14}} \begin{pmatrix} 1 \\ 2 \\ 3 \end{pmatrix}$ **b)** $\mathbf{q}_2 = \frac{1}{\sqrt{3}} \begin{pmatrix} -1 \\ -1 \\ 1 \end{pmatrix}$ **c)** $\mathbf{q}_3 = \frac{1}{\sqrt{42}} \begin{pmatrix} 5 \\ -4 \\ 1 \end{pmatrix}$.

10. **a)** $\mathbf{q}_1 = \frac{1}{\sqrt{14}} \begin{pmatrix} 1 \\ 2 \\ 3 \end{pmatrix}$ **b)** $\mathbf{q}_2 = \frac{1}{\sqrt{3}} \begin{pmatrix} -1 \\ -1 \\ 1 \end{pmatrix}$ **c)** $\mathbf{q}_3 = \frac{1}{\sqrt{42}} \begin{pmatrix} 5 \\ -4 \\ 1 \end{pmatrix}$.

11. **a)** $-1: \begin{pmatrix} -1 \\ 1 \\ 3 \end{pmatrix} \quad 1: \begin{pmatrix} 1 \\ 1 \\ 1 \end{pmatrix} \quad 2: \begin{pmatrix} 4 \\ 2 \\ 3 \end{pmatrix}$

b) $-2: \begin{pmatrix} -1 \\ 2 \\ 1 \end{pmatrix} \quad 1: \begin{pmatrix} -1 \\ -1 \\ 1 \end{pmatrix} \quad 2: \begin{pmatrix} 1 \\ 0 \\ 1 \end{pmatrix}$.

12. **a)** $-16: \begin{pmatrix} -1 \\ -2 \\ 2 \end{pmatrix} \quad 2: \begin{pmatrix} -2 \\ 1 \\ 0 \end{pmatrix} \quad 2: \begin{pmatrix} -2 \\ -4 \\ -5 \end{pmatrix}$

b) $6: \begin{pmatrix} 1 \\ 1 \\ 1 \end{pmatrix} \quad 3: \begin{pmatrix} -1 \\ 1 \\ 0 \end{pmatrix} \quad 3: \begin{pmatrix} -1 \\ -1 \\ 2 \end{pmatrix}$.

18. $\begin{pmatrix} 0 \\ 1 \\ -2 \end{pmatrix} \quad \begin{pmatrix} 1 \\ 0 \\ 0 \end{pmatrix} \quad \begin{pmatrix} 0 \\ 2 \\ 1 \end{pmatrix}$.

19. $\begin{pmatrix} 2 \\ 1 \\ 0 \end{pmatrix} \quad \begin{pmatrix} -1 \\ 2 \\ 5 \end{pmatrix} \quad \begin{pmatrix} 1 \\ -2 \\ 1 \end{pmatrix}$.

20. $\begin{pmatrix} 1 \\ 0 \\ -1 \end{pmatrix} \quad \begin{pmatrix} 1 \\ 4 \\ 1 \end{pmatrix} \quad \begin{pmatrix} 2 \\ -1 \\ 2 \end{pmatrix}$.

21. **a)** $\mathbf{q}_2 = \frac{1}{\sqrt{38}} \begin{pmatrix} 3 \\ -2 \\ 5 \end{pmatrix}$ **b)** $\lambda_2 = -18$ **c)** $\mathbf{q}_1 = \frac{1}{\sqrt{114}} \begin{pmatrix} 7 \\ 8 \\ -1 \end{pmatrix}$.

22. **a)** $\mathbf{q}_1 = \frac{1}{\sqrt{6}} \begin{pmatrix} 1 \\ -1 \\ 2 \end{pmatrix}$ **b)** $\lambda_1 = 8$ **c)** $\mathbf{q}_2 = \frac{1}{\sqrt{2}} \begin{pmatrix} 1 \\ 1 \\ 0 \end{pmatrix}$.

27. $\lambda = 0$.

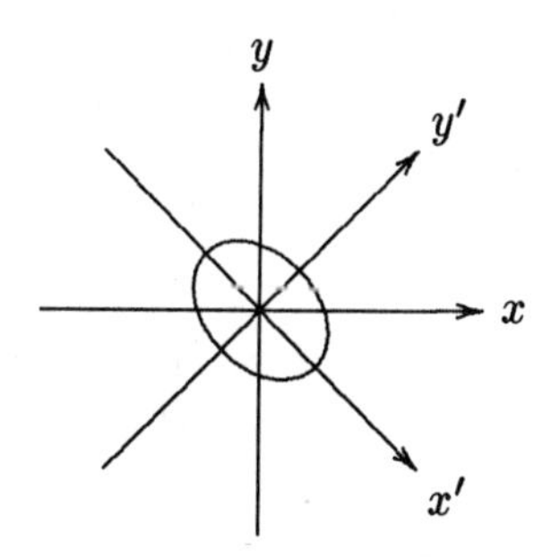

Figure A.4

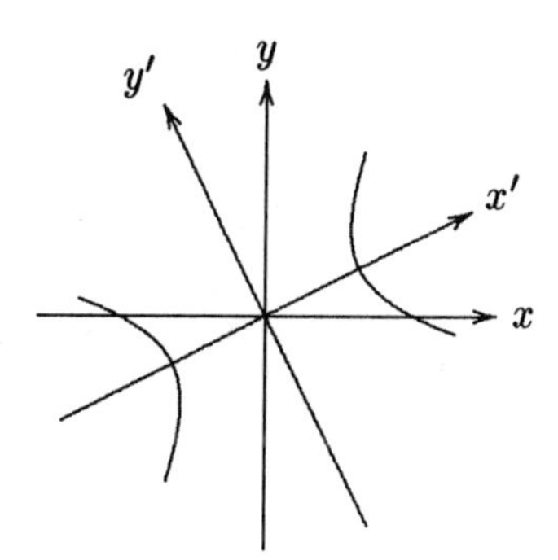

Figure A.5

A.8 Chapter 8

1. **a)** $\begin{pmatrix} 13 & -6 \\ -6 & 19 \end{pmatrix}$ **b)** $\begin{pmatrix} 6 & 1 \\ 1 & 14 \end{pmatrix}$ **c)** $\begin{pmatrix} 4 & 9 \\ 9 & 4 \end{pmatrix}$.

2. **a)** $\mathbf{q}_1 = \frac{1}{\sqrt{13}} \begin{pmatrix} 2 \\ 3 \end{pmatrix}, \quad \mathbf{q}_2 = \frac{1}{\sqrt{13}} \begin{pmatrix} -3 \\ 2 \end{pmatrix}, \quad 5x'^2 - 8y'^2 = 1$

b) $\mathbf{q}_1 = \frac{1}{\sqrt{29}} \begin{pmatrix} 5 \\ -2 \end{pmatrix}, \quad \mathbf{q}_2 = \frac{1}{\sqrt{29}} \begin{pmatrix} 2 \\ 5 \end{pmatrix}, \quad 9x'^2 - 20y'^2 = 1$

c) $\mathbf{q}_1 = \frac{1}{5} \begin{pmatrix} 3 \\ 4 \end{pmatrix}, \quad \mathbf{q}_2 = \frac{1}{5} \begin{pmatrix} -4 \\ 3 \end{pmatrix}, \quad 13x'^2 - 12y'^2 = 1$.

3. **a)** $2x'^2 + 4y'^2 = 16$ ellipse $x'-$axis : $\langle 1, -1 \rangle$ $y'-$axis : $\langle 1, 1 \rangle$. See Figure A.4.

b) $2x'^2 - 3y'^2 = 6$ hyperbola $x'-$axis: $\langle 2, 1 \rangle$ $y'-$axis: $\langle -1, 2 \rangle$. See Figure A.5.

c) $(x' - 1)^2 + 3y'^2 = 1$ ellipse center at $x' = 1$, $y' = 0$ $x'-$axis: $\langle 1, -1 \rangle$ $y'-$axis: $\langle 1, 1 \rangle$. See Figure A.6.

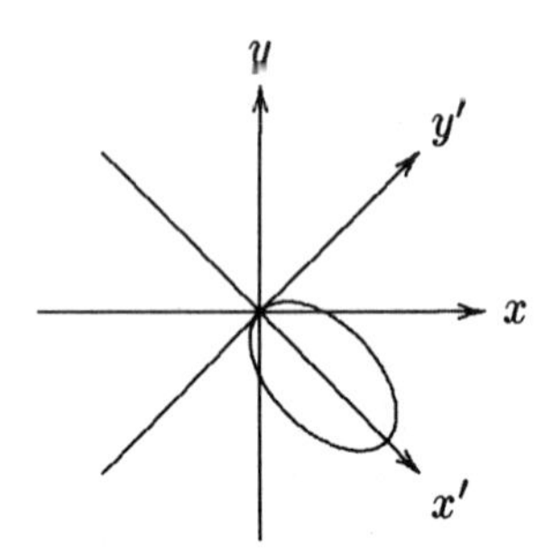

Figure A.6

4. **a)** The two eigenvalues are distinct, but have the same sign.
   b) The two eigenvalues have opposite signs.
   c) Exactly one of the eigenvalues is zero.

5. **a)** $\begin{pmatrix} 13 & -12 & 4 \\ -12 & -29 & 17 \\ 4 & 17 & 41 \end{pmatrix}$ **b)** $\begin{pmatrix} 2 & -4 & 8 \\ -4 & 5 & 14 \\ 8 & 14 & 3 \end{pmatrix}$ **c)** $\begin{pmatrix} 3 & 1 & 16 \\ 1 & 8 & 13 \\ 16 & 13 & 10 \end{pmatrix}$

   d) $\begin{pmatrix} 9 & 7 & 11 \\ 7 & 16 & 17 \\ 11 & 17 & 25 \end{pmatrix}$.

6. **a)** $\mathbf{q}_1 = \frac{1}{\sqrt{3}}\begin{pmatrix} 1 \\ 1 \\ 1 \end{pmatrix}$, $\mathbf{q}_2 = \frac{1}{\sqrt{6}}\begin{pmatrix} -2 \\ 1 \\ 1 \end{pmatrix}$, and $\mathbf{q}_3 = \frac{1}{\sqrt{2}}\begin{pmatrix} 0 \\ -1 \\ 1 \end{pmatrix}$

   b) $Q = \begin{pmatrix} \frac{1}{\sqrt{3}} & \frac{-2}{\sqrt{6}} & 0 \\ \frac{1}{\sqrt{3}} & \frac{1}{\sqrt{6}} & \frac{-1}{\sqrt{2}} \\ \frac{1}{\sqrt{3}} & \frac{1}{\sqrt{6}} & \frac{1}{\sqrt{2}} \end{pmatrix}$

   c) $3x'^2 + 6y'^2 - 2z'^2 = 100$.

7. **a)** $\mathbf{q}_1 = \frac{1}{\sqrt{42}}\begin{pmatrix} -5 \\ 4 \\ 1 \end{pmatrix}$, $\mathbf{q}_2 = \frac{1}{\sqrt{3}}\begin{pmatrix} 1 \\ 1 \\ 1 \end{pmatrix}$, and $\mathbf{q}_3 = \frac{1}{\sqrt{14}}\begin{pmatrix} 1 \\ 2 \\ -3 \end{pmatrix}$

   b) $Q = \begin{pmatrix} \frac{-5}{\sqrt{42}} & \frac{1}{\sqrt{3}} & \frac{1}{\sqrt{14}} \\ \frac{4}{\sqrt{42}} & \frac{1}{\sqrt{3}} & \frac{2}{\sqrt{14}} \\ \frac{1}{\sqrt{42}} & \frac{1}{\sqrt{3}} & \frac{-3}{\sqrt{14}} \end{pmatrix}$

c) $x'^2 - 2y'^2 + 15z'^2 = 100$.

8. **a)** $\mathbf{p}_1 = \begin{pmatrix} 1 \\ 1 \\ -1 \end{pmatrix}$

b) $\lambda_2 = -2$ and $\lambda_3 = 12$

c) $\mathbf{q}_1 = \frac{1}{\sqrt{3}} \begin{pmatrix} 1 \\ 1 \\ -1 \end{pmatrix}$, $\mathbf{q}_2 = \frac{1}{\sqrt{14}} \begin{pmatrix} 1 \\ 2 \\ 3 \end{pmatrix}$, $\mathbf{q}_3 = \frac{1}{\sqrt{42}} \begin{pmatrix} 5 \\ -4 \\ 1 \end{pmatrix}$ hence

$$Q = \begin{pmatrix} \frac{1}{\sqrt{3}} & \frac{1}{\sqrt{14}} & \frac{5}{\sqrt{42}} \\ \frac{1}{\sqrt{3}} & \frac{2}{\sqrt{14}} & \frac{-4}{\sqrt{42}} \\ \frac{-1}{\sqrt{3}} & \frac{3}{\sqrt{14}} & \frac{1}{\sqrt{42}} \end{pmatrix}$$

d) $9x'^2 - 2y'^2 + 12z'^2 = 1$.

9. **a)** $\mathbf{p}_1 = \begin{pmatrix} 2 \\ 2 \\ -3 \end{pmatrix}$

b) $\lambda_2 = 3$ and $\lambda_3 = -6$

c) $\mathbf{q}_1 = \frac{1}{\sqrt{17}} \begin{pmatrix} 2 \\ 2 \\ -3 \end{pmatrix}$, $\mathbf{q}_2 = \frac{1}{3} \begin{pmatrix} 1 \\ 2 \\ 2 \end{pmatrix}$, $\mathbf{q}_3 = \frac{1}{3\sqrt{17}} \begin{pmatrix} 10 \\ -7 \\ 2 \end{pmatrix}$ hence

$$Q = \begin{pmatrix} \frac{2}{\sqrt{17}} & \frac{1}{3} & \frac{10}{3\sqrt{17}} \\ \frac{2}{\sqrt{17}} & \frac{2}{3} & \frac{-7}{3\sqrt{17}} \\ \frac{-3}{\sqrt{17}} & \frac{2}{3} & \frac{2}{3\sqrt{17}} \end{pmatrix}$$

d) $11x'^2 + 3y'^2 - 6z'^2 = 1$.

10. Let $\mathbf{q}_1 = \frac{1}{2}\mathbf{p}_2$, $\mathbf{q}_2 = \mathbf{p}_1$, $\mathbf{q}_3 = \frac{1}{3}\mathbf{p}_3$, and $Q = [\mathbf{q}_1, \mathbf{q}_2, \mathbf{q}_3]$.

11. **a)** $\mathbf{p}_3 = \begin{pmatrix} 1 \\ 3 \\ -5 \end{pmatrix}$ and $\lambda_3 = 29$ **b)** $\mathbf{p}_2 = \begin{pmatrix} 13 \\ 4 \\ 5 \end{pmatrix}$ **c)** $-6x'^2 - 6y'^2 + 29z'^2 = 1$.

12. **a)** x'– axis: $\mathbf{q}_1 = \frac{1}{\sqrt{3}}\begin{pmatrix} 1 \\ -1 \\ 1 \end{pmatrix}$ y'– axis: $\mathbf{q}_2 = \frac{1}{\sqrt{2}}\begin{pmatrix} 1 \\ 1 \\ 0 \end{pmatrix}$ z'– axis: $\mathbf{q}_3 = \frac{1}{\sqrt{6}}\begin{pmatrix} -1 \\ 1 \\ 2 \end{pmatrix}$.

b) The equation in the new coordinates is

$$6x'^2 + 6y'^2 + 12z'^2 = 24$$

which is an ellipsoid.

13. **a)** The principal axes lie along the vectors $\mathbf{q}_1$, $\mathbf{q}_2$, and $\mathbf{q}_3$ where $\mathbf{q}_1 = \frac{1}{9}\begin{pmatrix} -4 \\ 8 \\ 1 \end{pmatrix}$, $\mathbf{q}_2 = \frac{1}{9}\begin{pmatrix} -7 \\ -4 \\ 4 \end{pmatrix}$, and $\mathbf{q}_3 = \frac{1}{9}\begin{pmatrix} 4 \\ 1 \\ 8 \end{pmatrix}$.

b)

$$-8x'^2 + y'^2 + 10z'^2 + 2x' + 5y' + 4z' = 48 \quad .$$

14. **a)** $A\mathbf{w}_1 = -4\mathbf{w}_1$, $A\mathbf{w}_2 = -\mathbf{w}_2$, $A\mathbf{w}_3 = 2\mathbf{w}_3$, $B\mathbf{w}_1 = 2\mathbf{w}_1$, $B\mathbf{w}_2 = 5\mathbf{w}_2$, $B\mathbf{w}_3 = -\mathbf{w}_3$.

b) $-4x'^2 - y'^2 + 2z'^2 = 40$, $2x'^2 + 5y'^2 - z'^2 = 40$, and $2x'^2 + 5y'^2 - z'^2 = 0$.

c) See Figures A.7, A.8, and A.9.

15. $-8x'^2 - 5y'^2 - 2z'^2 - x' + y' + z' = -26$.

A.9 Chapter 9

1. $\begin{pmatrix} -1 & 3 \\ -2 & 0 \end{pmatrix}$.

2. **a)** $x' = 2$, $y' = -3$, and $z' = -2$.

b) $x = 2$, $y = -2$, and $z = -1$.

3. $X = \begin{pmatrix} -7 & 2 & -18 \\ 5 & -1 & 13 \end{pmatrix}$.

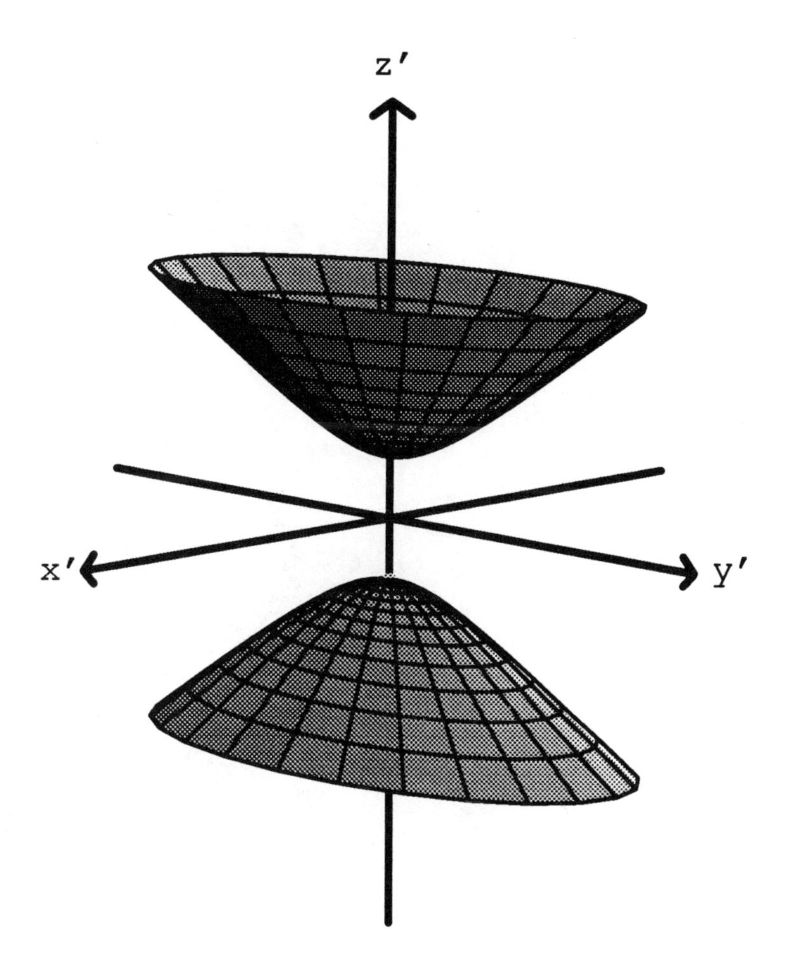

Figure A.7 Hyperboloid of two sheets

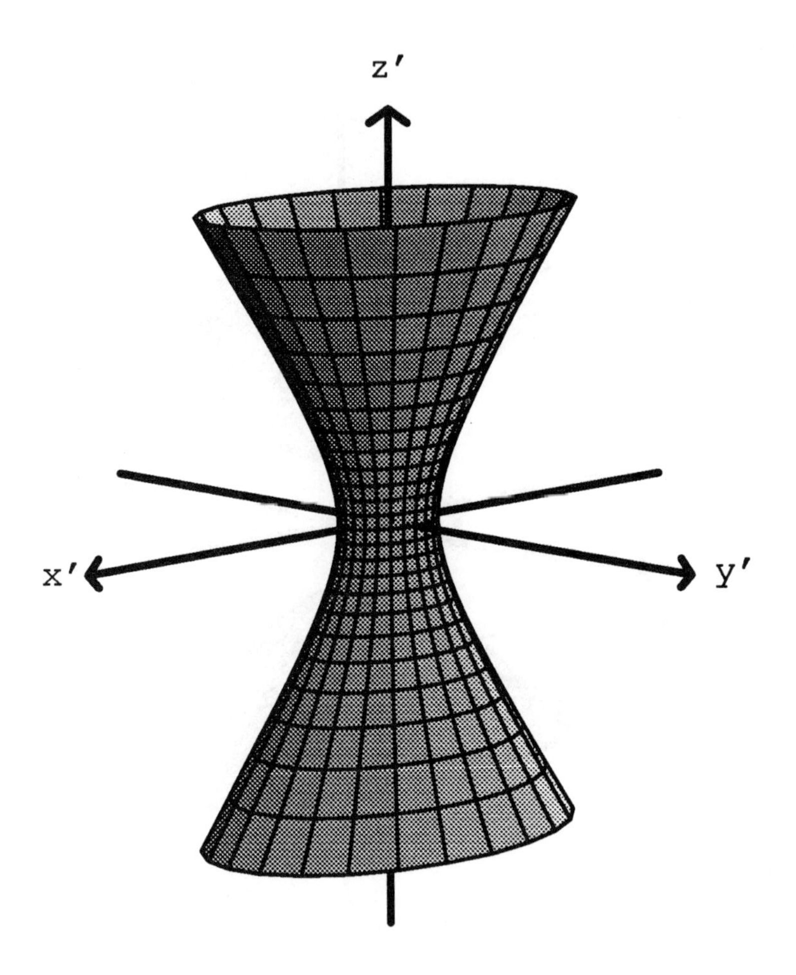

Figure A.8 Hyperboloid of one sheet

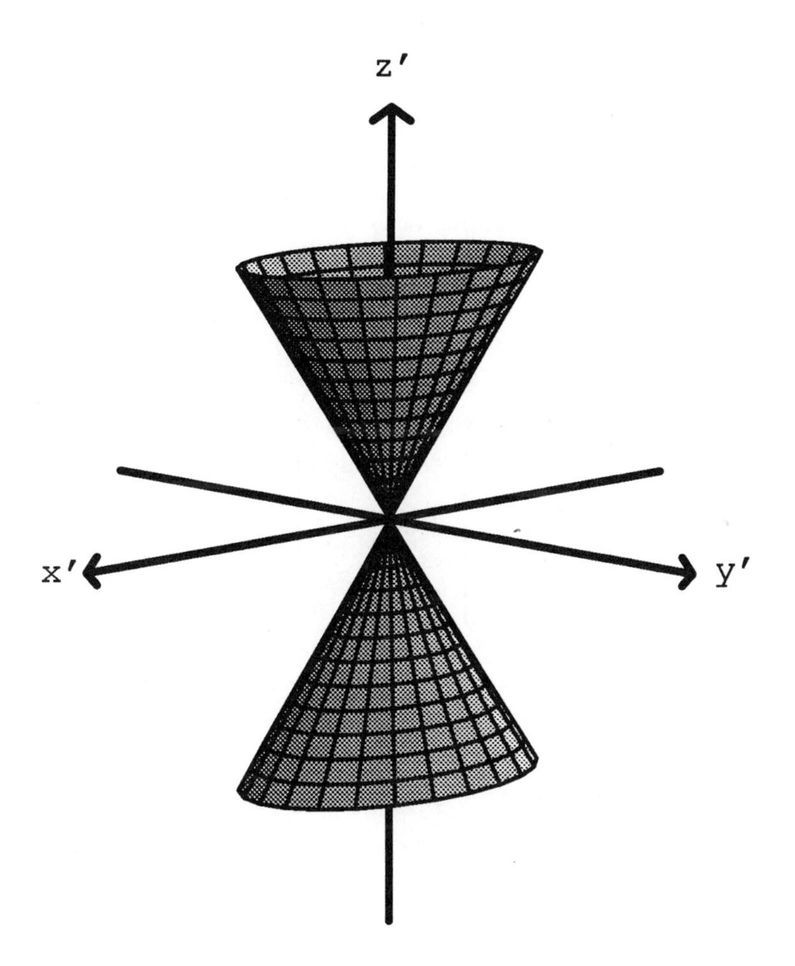

Figure A.9 Elliptic Cone

4. There is no solution

5. $X = \begin{pmatrix} 9 & 1 \\ 1 & -3 \\ 0 & -14 \end{pmatrix}$.

6. **a)** $\begin{pmatrix} 0 & 0 \\ 0 & 0 \end{pmatrix}$ **b)** $\begin{pmatrix} 10 & 20 \\ -5 & -10 \end{pmatrix}$ **c)** $\begin{pmatrix} -1 & 11 & -5 \\ 2 & -22 & 10 \end{pmatrix}$

d) Incompatible Dimensions **e)** $\begin{pmatrix} 0 & 8 \\ 7 & 10 \end{pmatrix}$ **f)** Incompatible Dimensions.

7. 2.

8. $X = \begin{pmatrix} 1 & 2 & -3 \\ 1 & 3 & -4 \\ -5 & -12 & 18 \end{pmatrix}$.

9. **a)** 1 **b)** -18 **c)** -12 **d)** 60 **e)** -24

10. $X = \begin{pmatrix} 0 \\ 8 \\ 0 \\ 6 \\ 7 \\ 0 \end{pmatrix} + a \begin{pmatrix} 1 \\ 0 \\ 0 \\ 0 \\ 0 \\ 0 \end{pmatrix} + b \begin{pmatrix} 0 \\ 3 \\ 1 \\ 0 \\ 0 \\ 0 \end{pmatrix} + c \begin{pmatrix} 0 \\ -2 \\ 0 \\ 5 \\ -4 \\ 1 \end{pmatrix}$.

11. $\begin{pmatrix} 1 & -1 & 0 \\ -1 & 2 & -1 \\ 0 & -1 & 2 \end{pmatrix}$.

12. $X = \begin{pmatrix} 3 \\ -2 \\ 4 \end{pmatrix}$.

13. **a)** $\begin{pmatrix} 1 & 2 & -1 \end{pmatrix}, \left(\begin{array}{ccc|c} 1 & 2 & -1 & 3 \end{array}\right)$ **b)** $\begin{pmatrix} 1 & 3 & 5 \\ 2 & 4 & 6 \end{pmatrix}, \left(\begin{array}{ccc|c} 1 & 3 & 5 & 7 \\ 2 & 4 & 6 & 8 \end{array}\right)$

c) $\begin{pmatrix} 1 & 1 \\ 2 & 1 \\ 3 & 1 \end{pmatrix}, \left(\begin{array}{cc|c} 1 & 1 & 0 \\ 2 & 1 & 1 \\ 3 & 1 & 2 \end{array}\right)$ **d)** $\begin{pmatrix} 1 & 1 & 1 \\ 1 & -1 & 1 \\ 1 & -1 & -1 \end{pmatrix}, \left(\begin{array}{ccc|c} 1 & 1 & 1 & 0 \\ 1 & -1 & 1 & 1 \\ 1 & -1 & -1 & 2 \end{array}\right)$.

14. **a)** $2x_1 = 3$ **b)** $\begin{aligned} 2x_1 + x_2 &= 0 \\ 3x_1 + 4x_2 &= 5 \end{aligned}$ **c)** $\begin{aligned} x_1 + 3x_2 &= 1 \\ 2x_1 + x_2 &= 2 \\ 3x_1 + 2x_2 &= 1 \end{aligned}$.

15. a) $\begin{matrix} x = 1 \\ y = 1 \end{matrix}$ b) $\begin{matrix} x = 3 \\ y = -2 \end{matrix}$ c) $\begin{matrix} x = -2 \\ y = \ \ 3 \end{matrix}$ d) $\begin{matrix} x = 0 \\ y = 0 \end{matrix}$.

16. a) $\begin{matrix} x = \ \ 3 \\ y = -2 \end{matrix}$ b) No solution c) $\begin{matrix} x = 0 \\ y = 0 \end{matrix}$ d) No solution .

17. a) $\mathbf{x} = \begin{pmatrix} 3\alpha - 2\beta \\ 1 - 2\alpha \\ \alpha \\ \beta \end{pmatrix}$ b) $\mathbf{x} = \begin{pmatrix} 2t \\ -2t \\ -t \\ t \end{pmatrix}$ c) $\mathbf{x} = \begin{pmatrix} 2 - t \\ t \\ t \end{pmatrix}$ d) $\mathbf{x} = \begin{pmatrix} 2 \\ -2 \\ -1 \\ 3 \end{pmatrix}$

18. a) $\begin{pmatrix} 1 & 1 & 0 & 0 \\ 0 & 0 & 1 & 1 \end{pmatrix}$ b) $\begin{pmatrix} 1 & 3 & 0 & 1 & 0 & 2 \\ 0 & 0 & 1 & -2 & 0 & 1 \\ 0 & 0 & 0 & 0 & 1 & -1 \end{pmatrix}$

c) $\begin{pmatrix} 1 & 0 & 1 & -1 & 0 \\ 0 & 1 & 2 & -1 & 0 \\ 0 & 0 & 0 & 0 & 1 \end{pmatrix}$ d) $\begin{pmatrix} 1 & 0 & 1 \\ 0 & 1 & 1 \\ 0 & 0 & 0 \\ 0 & 0 & 0 \end{pmatrix}$.

19. a) -7 b) 13 c) -13 d) -7 e) -13 .

20. a) $\frac{1}{7}\begin{pmatrix} 5 & -2 \\ -4 & 3 \end{pmatrix}$ b) $\begin{pmatrix} 7 & -3 & 0 \\ -11 & 6 & -1 \\ 9 & -5 & 1 \end{pmatrix}$

c) $\begin{pmatrix} 1 & -5 & -12 & 8 \\ 0 & 1 & 2 & -1 \\ 0 & 0 & 1 & -2 \\ 0 & 0 & 0 & 1 \end{pmatrix}$ d) $\begin{pmatrix} 1 & 0 & -2 & 0 \\ 2 & 1 & -4 & -3 \\ 8 & 4 & -15 & -12 \\ 0 & 0 & 0 & 1 \end{pmatrix}$.

21. a) $\begin{pmatrix} -7 & -7 \\ \frac{11}{2} & 6 \end{pmatrix}$ b) $\begin{pmatrix} 2 & -6 \\ 7 & -7 \\ -5 & 9 \end{pmatrix}$ c) $\frac{-1}{7}\begin{pmatrix} 4 & -5 & -7 \\ -3 & 2 & 0 \end{pmatrix}$

d) $\begin{pmatrix} 2 & -9 \\ -2 & 1 \\ 1 & 2 \\ -1 & 1 \end{pmatrix}$.

22. a) $\frac{1}{7}\begin{pmatrix} 5 \\ 3 \end{pmatrix}, \frac{1}{7}\begin{pmatrix} 3 & 1 \\ -1 & 2 \end{pmatrix}$ b) $\begin{pmatrix} 5 \\ 0 \\ -11 \end{pmatrix}, \begin{pmatrix} 1 & 0 & -3 \\ -1 & 1 & 1 \\ -2 & 0 & 7 \end{pmatrix}$

c) $\begin{pmatrix} 5 \\ -2 \\ -4 \end{pmatrix}$, $\begin{pmatrix} 1 & 1 & -1 \\ 0 & -1 & 1 \\ -1 & 0 & 1 \end{pmatrix}$ **d)** $\begin{pmatrix} 3 \\ 4 \\ 4 \\ -17 \end{pmatrix}$, $\begin{pmatrix} 1 & 0 & 0 & -1 \\ 0 & 1 & -1 & 0 \\ -1 & 2 & -1 & 1 \\ -2 & -3 & 3 & 3 \end{pmatrix}$.

23. The given matrices can be row reduced to the following row-reduced echelon matrices:

a) $\begin{pmatrix} 1 & 2 \\ 0 & 0 \end{pmatrix}$ **b)** $\begin{pmatrix} 1 & 0 & -2 \\ 0 & 1 & -1 \\ 0 & 0 & 0 \end{pmatrix}$ **c)** $\begin{pmatrix} 1 & 0 & 0 & -2 \\ 0 & 1 & 0 & 3 \\ 0 & 0 & 1 & -1 \\ 0 & 0 & 0 & 0 \end{pmatrix}$

d) $\begin{pmatrix} 1 & 0 & 0 & 0 & 0 \\ 0 & 1 & 0 & 1 & 0 \\ 0 & 0 & 1 & 0 & 0 \\ 0 & 0 & 0 & 0 & 1 \\ 0 & 0 & 0 & 0 & 0 \end{pmatrix}$.

24. **a)** $(2,2,5)$ **b)** $(4,0,5)$ **c)** $(1,1,8)$.

25. **a)** $A = \begin{pmatrix} 1 & 2 & 1 \\ -1 & 1 & -1 \end{pmatrix}$ and $B = \begin{pmatrix} 2 & 1 \\ 1 & -2 \\ -1 & 1 \end{pmatrix}$

b) $M = \begin{pmatrix} 1 & 5 & 1 \\ 3 & 0 & 3 \\ -2 & -1 & -2 \end{pmatrix}$.

26. **a)** $\begin{pmatrix} 2 & 8 & 4 & 6 \\ 2 & -2 & 10 & -2 \end{pmatrix}$

b) $\begin{pmatrix} 5 & 34 & 11 \\ -4 & -14 & -4 \\ 15 & 36 & 9 \end{pmatrix}$

c) $\begin{pmatrix} 0 \end{pmatrix}$.

27. 4.

28. -14.

29.

$$A = \begin{pmatrix} 1 & 2 & 3 \\ 3 & 7 & 12 \\ 4 & 10 & 19 \end{pmatrix}.$$

30. $q_{12} = -2$, $q_{13} = 3$, $q_{23} = -2$, $q_{32} = 3$, and $q_{33} = 6$.

$$\therefore \qquad Q = \frac{1}{7}\begin{pmatrix} 6 & -2 & 3 \\ 3 & 6 & -2 \\ -2 & 3 & 6 \end{pmatrix} .$$

31.

$$X = \begin{pmatrix} 4 & -2 & 1 \\ 11 & 0 & 1 \\ -5 & 5 & -2 \end{pmatrix} .$$

32. a)

$$S - 2I = \begin{pmatrix} 1 & -1 & 2 \\ -1 & 1 & -2 \\ 2 & -2 & 4 \end{pmatrix} \sim \begin{pmatrix} 1 & -1 & 2 \\ 0 & 0 & 0 \\ 0 & 0 & 0 \end{pmatrix}$$

$x-y+2z = 0$ is the equation for the eigenvectors corresponding to $\lambda_3 = \lambda_1 = 2$.

$$\mathbf{v}_2 = \begin{pmatrix} 1 \\ -1 \\ 2 \end{pmatrix}$$

is a normal vector to this plane and is another eigenvector. $S\mathbf{v}_2 = 8\mathbf{v}_2$ so $\lambda_2 = 8$.

b)

$$\mathbf{v}_2 \times \mathbf{v}_3 = \mathbf{v}_1 = \begin{pmatrix} -3 \\ -3 \\ 0 \end{pmatrix}$$

is perpendicular to $\mathbf{v}_3$ and hence corresponds to $\lambda_1 = 2$.

A.10 Chapter 10

1. **a)** $e^{x^2y^3}$ **b)** $\operatorname{sech}^2(x^2 + y^3 + z)$ **c)** $\dfrac{1}{x^2 + y^4}$.

2. **a)** $\begin{pmatrix} \cos u_2 & -u_1 \sin u_2 & 0 \\ \cos u_2 & -u_1 \sin u_2 & 0 \\ 0 & 0 & 1 \end{pmatrix}$

b) $\begin{pmatrix} \sin u_2 \cos u_3 & u_1 \cos u_2 \cos u_3 & u_1 \sin u_2 \sin u_3 \\ \sin u_2 \sin u_3 & u_1 \cos u_2 \sin u_3 & u_1 \cos u_3 \sin u_2 \\ \cos u_2 & -u_1 \sin u_2 & 0 \end{pmatrix}$

c) $\begin{pmatrix} -a \sin u_1 \cos u_2 & -a \cos u_1 \sin u_2 \\ a \cos u_1 \cos u_2 & -a \sin u_1 \sin u_2 \\ 0 & a \cos u_2 \end{pmatrix}$ **d)** $\begin{pmatrix} 2u_1 & -2u_2 \\ 2u_2 & 2u_1 \end{pmatrix}$ **e)** $\begin{pmatrix} 2 & 3 \\ 1 & 1 \end{pmatrix}$

f) $\begin{pmatrix} 2 & 1 & -1 \\ -1 & 2 & 3 \end{pmatrix}$ **g)** $\begin{pmatrix} -1 & 2 & 3 \\ 1 & 1 & -1 \\ 2 & -3 & 2 \end{pmatrix}$.

3. **a)** $\langle -u_1 + u_2, -u_1 - 4u_2 \rangle$ **b)** $\langle 4u_1 - 2u_2 - 4u_3, 2u_1 + 14u_2 + 2u_3, 8u_1 + 20u_2 - 2u_3 \rangle$
c) $\langle -3u_1 + u_2 - 13u_3, -u_1 - 4u_2 - 13u_3, 0 \rangle$.

A.11 Chapter 11

1. **a)** $3 - 3(x-1) + (y-2) + 3(x-1)^2 - (x-1)(y-2)$
b) $(y+1) - (x-1)(y+1)$
c) $1 - \frac{1}{3}(x-3) + \frac{1}{3}(y-2) + \frac{1}{9}(x-3)^2 - \frac{1}{9}(x-3)(y-2)$
d) $\frac{1}{3}(y+1) - \frac{1}{9}(x-3)(y+1)$.

2. **a)** $361,\ \langle -19, 18 \rangle,\ \begin{pmatrix} 26 & 3 \\ 3 & 14 \end{pmatrix}$

b) $381,\ \langle 12, 27 \rangle,\ \begin{pmatrix} 34 & -1 \\ -1 & 2 \end{pmatrix}$

c) $339,\ \langle 2, 11 \rangle,\ \begin{pmatrix} 30 & 11 \\ 11 & 14 \end{pmatrix}$

d) $417,\ \langle -44, 33 \rangle,\ \begin{pmatrix} 22 & -5 \\ -5 & 14 \end{pmatrix}$.

3. **a)** $121,\ \langle 9, 7 \rangle,\ \begin{pmatrix} 10 & 4 \\ 4 & -6 \end{pmatrix}$

b) $356,\ \langle 236, -90 \rangle,\ \begin{pmatrix} 58 & -110 \\ -110 & 0 \end{pmatrix}$.

4. **a)** $1 + (x-3) - (y-2) - (z-1) + (z-1)^2 - (x-3)(z-1) + (y-2)(z-1)$
b) $-1 + (x-2) - (y-3) + (z-1) - (z-1)^2 - (x-2)(z-1) + (y-3)(z-1)$
c) $1 - (x-1) + (y-2) + (z+1) + (z+1)^2 - (x-1)(z+1) + (y-2)(z+1)$

d) $2+(x-1)-(y+1)-2(z-1)+2(z-1)^2-(x-1)(z-1)+(y+1)(z-1)$

.

5. a) $\langle 3,-6,4\rangle,\ \begin{pmatrix} -2 & 2 & -2 \\ 2 & -14 & 12 \\ -2 & 12 & -6 \end{pmatrix},$

$-3+3(x+1)-6(y-1)+4(z+1)-(x+1)^2-7(y-1)^2-3(z+1)^2+$
$2(x+1)(y-1)-2(x+1)(z+1)+12(y-1)(z+1)$

b) $\langle -1,2,4\rangle,\ \begin{pmatrix} 2 & 2 & -2 \\ 2 & 10 & 12 \\ -2 & 12 & 6 \end{pmatrix},$

$1-(x+1)+2(y-1)+4(z-1)+(x+1)^2+5(y-1)^2+$
$3(z-1)^2+2(x+1)(y-1)-2(x+1)(z-1)+12(y-1)(z-1)$

c) $\langle -1,2,4\rangle,\ \begin{pmatrix} -2 & -2 & 2 \\ -2 & -10 & -12 \\ 2 & -12 & -6 \end{pmatrix},$

$-1-(x-1)+2(y+1)+4(z+1)-(x-1)^2-5(y+1)^2-$
$3(z+1)^2-2(x-1)(y+1)+2(x-1)(z+1)-12(y+1)(z+1)$

d) $\langle -1,-2,4\rangle\ \begin{pmatrix} -2 & 2 & 2 \\ 2 & -10 & 12 \\ 2 & 12 & -6 \end{pmatrix}$

$-1-(x-1)-2(y-1)+4(z+1)-(x-1)^2-5(y-1)^2-$
$3(z+1)^2+2(x-1)(y-1)+2(x-1)(z+1)+12(y-1)(z+1)$.

6. a) $2(x-1)+2(y-1)+2(z+1)-(x-1)^2-2(y-1)^2-$
$2(z-1)^2-4(x-1)(y-1)-4(x-1)(z+1)-4(y-1)(z+1)$

b) $\frac{-\pi}{2}-(x-1)-(y+1)-\frac{\pi}{2}(z-1)+$
$\frac{1}{2}(x-1)^2-\frac{1}{2}(y+1)^2-(x-1)(z-1)-(y+1)(z-1)$

c) $-3+4(x-1)+3(y+1)-15(z-1)+$
$10(x-1)^2-3(y+1)^2-30(z-1)^2+3(x-1)(y+1)$

d) $2-3(x-1)+6(y+1)-3(z+1)$
$+5(x-1)(y+1)+5(x-1)(z+1)-3(y+1)(z+1)$

7. a) $\langle -3,2,0\rangle,\ \begin{pmatrix} 4 & 8 & 10 \\ 8 & 6 & 6 \\ 10 & 6 & -10 \end{pmatrix}$

b) $3-3x+2y+2x^2+3y^2-5z^2+8xy+10xz+6yz$.

8. **a)** $\langle 4,0,-1\rangle \quad \begin{pmatrix} 6 & 16 & 14 \\ 16 & 2 & 8 \\ 14 & 8 & 14 \end{pmatrix}$

b) $5+4x-z+3x^2+y^2+7z^2+16xy+14xz+8yz$.

A.12 Chapter 12

1. **a)** $(-2,3)$ local minimum **b)** $(5,-3)$ local maximum **c)** $(1,-1)$ local maximum

d) $(2,2)$ local minimum, $(-2,2)$ saddle point, $(0,2)$ saddle point, $(2,0)$ saddle point, $(-2,0)$ local maximum, $(0,0)$ local minimum.

2. **a)** $(1,-1,-1)$ local minimum **b)** $(-2,-1,1)$ local minimum **c)** $(-\frac{1}{3},\frac{2}{3},7)$ local minimum **d)** $(-\frac{2}{3},-\frac{1}{3},\frac{1}{2})$ local minimum

e) $(2,1,-2)$ local maximum **f)** $(1,3,1)$ saddle point, $(1,3,-1)$ local maximum **g)** $(-2,-1,2)$ saddle point, $(-2,3,2)$ saddle point, $(1,-1,2)$ saddle point, $(1,3,2)$ local minimum.

h) $(-2,3,1)$ local minimum **i)** $(3,3,-2)$ local minimum, $(3,-3,-2)$ local minimum, $(3,0,-2)$ saddle point, $(0,3,-2)$ saddle point, $(0,-3,-2)$ saddle point, $(0,0,-2)$ saddle point.

3. **a)** $c<-1$ **b)** $c>1$ **c)** $-1<c<0$ or $0<c<1$ **d)** $c=-1$, $c=0$, or $c=1$.

4. $(1,1,2)$ local minimum. $(2,2,-1)$ saddle point.

5. $(-1,2,-1)$ local minimum, $(2,2,-1)$ saddle point.

6. **a)** For A $\lambda_1=1$, $\lambda_2=6$, and $\lambda_3=9$ so it is a local minimum. For B $\lambda_1=-9$, $\lambda_2=-4$, and $\lambda_3=-1$ so it is a local maximum

b) For A $\lambda_1=1$, $\lambda_2=-4$, and $\lambda_3=-7$ so it is a saddle point. For B $\lambda_1=11$, $\lambda_2=6$, and $\lambda_3=3$ so it is a local minimum.

7. **a)** $(0,0,0)$ is a local minimum, $(1,1,1)$ is a saddle point, and the other critical points are $(-1,-1,1)$, $(-1,1,-1)$, $(1,-1,-1)$.

b) $(0,0,0)$ is a local maximum, $(1,1,1)$ is a saddle point, and the other critical points are $(-1,-1,1)$, $(-1,1,-1)$, $(1,-1,-1)$.

c) $(0,0,0)$ is a local minimum, $(2,2,2)$ is a saddle point, and the other critical points are $(-2,-2,2)$, $(-2,2,-2)$, $(2,-2,-2)$.

d) $(0,0,0)$ is a local maximum, $(\sqrt{3},1,1)$ is a saddle point, and the other critical points are $(-\sqrt{3},-1,1)$, $(-\sqrt{3},1,-1)$, $(\sqrt{3},-1,-1)$.

e) $(0,0,0)$ is a local minimum, $(2\sqrt{3},2,2)$ is a saddle point, and the other critical points are $(-2\sqrt{3},-2,2)$, $(-2\sqrt{3},2,-2)$, $(2\sqrt{3},-2,-2)$.

f) $(0,0,0)$ is a local maximum, $(2\sqrt{6},2,2)$ is a saddle point, and the other critical points are $(-2\sqrt{6},-2,2)$, $(-2\sqrt{6},2,-2)$, $(2\sqrt{6},-2,-2)$.

g) $(0,0,0)$ is a local minimum, $(\sqrt{6},1,1)$ is a saddle point, and the other critical points are $(-\sqrt{6},-1,1)$, $(-\sqrt{6},1,-1)$, $(\sqrt{6},-1,-1)$.

h) $(0,0,0)$ is a local minimum, $(\sqrt{3},1,1)$ is a saddle point, and the other critical points are $(-\sqrt{3},-1,1)$, $(-\sqrt{3},1,-1)$, $(\sqrt{3},-1,-1)$.

8. **a)** $(0,0,0)$ is a local minimum, $(-1,-1,-1)$ is a saddle point, and the other critical points are $(-1,1,1)$, $(1,-1,1)$, $(1,1,-1)$.

b) $(0,0,0)$ is a local maximum, $(-1,-1,-1)$ is a saddle point, and the other critical points are $(-1,1,1)$, $(1,-1,1)$, $(1,1,-1)$.

c) $(0,0,0)$ is a local minimum, $(-2,-2,-2)$ is a saddle point, and the other critical points are $(-2,2,2)$, $(2,-2,2)$, $(2,2,-2)$.

d) $(0,0,0)$ is a local minimum, $(-\sqrt{3},-1,-1)$ is a saddle point, and the other critical points are $(-\sqrt{3},1,1)$, $(\sqrt{3},-1,1)$, $(\sqrt{3},1,-1)$.

e) $(0,0,0)$ is a local maximum, $(-2\sqrt{3},-2,-2)$ is a saddle point, and the other critical points are $(-2\sqrt{3},2,2)$, $(2\sqrt{3},-2,2)$, $(2\sqrt{3},2,-2)$.

f) $(0,0,0)$ is a local minimum, $(-2\sqrt{6},-2,-2)$ is a saddle point, and the other critical points are $(-2\sqrt{6},2,2)$, $(2\sqrt{6},-2,2)$, $(2\sqrt{6},2,-2)$.

g) $(0,0,0)$ is a local maximum, $(-\sqrt{6},-1,-1)$ is a saddle point, and the other critical points are $(-\sqrt{6},1,1)$, $(\sqrt{6},-1,1)$, $(\sqrt{6},1,-1)$.

h) $(0,0,0)$ is a local minimum, $(-\sqrt{3},-1,-1)$ is a saddle point, and the other critical points are $(-\sqrt{3},1,1)$, $(\sqrt{3},-1,1)$, $(\sqrt{3},1,-1)$.

9. **a)** $(0,0,0)$ saddle point **b)** $(0,0,0)$ saddle point **c)** $(0,0,0)$ saddle point **d)** $(0,0,0)$ saddle point.

10. Hint: $\mathbf{Q}f = \mathbf{Q}g = \mathbf{Q}h = x^2 + y^2$.

A.13 Chapter 13

1. **a)** $(4,-2)$, $(1,4)$ **b)** $(3,-2)$, $(2,3)$

c) $(2,-1)$, $(1,2)$ **d)** $(1,1)$, $(1,-1)$.

2. **a)** $(2,-1)$, $(2,1)$ **b)** $\sqrt{3}(1,-1)$, $(1,1)$ **c)** $\frac{\sqrt{5}}{2}(3,1)$
d) $(2,-5)$ **e)** $(2,-2)$.

3. **a)** $\sqrt{3}(1,-1,0)$, $\sqrt{2}(1,1,1)$, $(1,1,-2)$
b) $\sqrt{15}(1,0,-1)$, $\sqrt{10}(1,1,1)$, $\sqrt{5}(1,-2,1)$
c) $\sqrt{2}(1,2,1)$, $\sqrt{6}(1,0,-1)$, $(2,-2,2)$
d) $(3,0,-3)$, $(1,4,1)$, $\sqrt{2}(2,-1,2)$.

4. **a)** $\sqrt{\frac{8}{3}}(1,0,-1)$, $\sqrt{2}(1,-1,0)$, $(1,1,2)$ **b)** $(2,0,0)$.

B Index